Data Analytics for Finance Using Python

Unlock the power of data analytics in finance with this comprehensive guide. *Data Analytics for Finance Using Python* is your key to unlocking the secrets of the financial markets.

In this book, you'll discover how to harness the latest data analytics techniques, including machine learning and inferential statistics, to make informed investment decisions and drive business success. With a focus on practical application, this book takes you on a journey from the basics of data preprocessing and visualization to advanced modeling techniques for stock price prediction.

Through real-world case studies and examples, you'll learn how to:

- Uncover hidden patterns and trends in financial data
- Build predictive models that drive investment decisions
- Optimize portfolio performance using data-driven insights
- Stay ahead of the competition with cutting-edge data analytics techniques

Whether you're a finance professional seeking to enhance your data analytics skills or a researcher looking to advance the field of finance through data-driven insights, this book is an essential resource. Dive into the world of data analytics in finance and discover the power to make informed decisions, drive business success, and stay ahead of the curve.

This book will be helpful for students, researchers, and users of machine learning and financial tools in the disciplines of commerce, management, and economics.

Advances in Digital Technologies for Smart Applications

Series Editor: Saad Motahhir

The Advances in Digital Technologies for Smart Applications series publishes leading-edge research on innovative digital technologies and their application in smart systems. Key topics include AI, IoT, blockchain, and their integration into various sectors, including finance, healthcare, and public governance.

Data Analytics for Finance Using Python
Nitin Jaglal Untwal, Utku Kose

Big Data and Blockchain Technology for Secure IoT Applications
Shitharth Selvarajan, Gouse Baig Mohammad, Sadda Bharath Reddy, Praveen Kumar Balachandran

Technology-Based Teaching and Learning in Pakistani English Language Classrooms
Muhammad Mooneeb Ali

Medical Knowledge Paradigms for Enabling the Digital Health Ecosystem
Usha Desai, Vivek P Chavda, Ankit Vijayvargiya, Ravichander Janapati

Soft Computing in Renewable Energy Technologies
Najib El Ouanjli, Mahmoud A. Mossa, Mariya Ouaissa, Sanjeevikumar Padmanaban, Said Mahfoud

Leveraging the Potential of Artificial Intelligence in the Real World: Smart Cities and Healthcare
Tien Anh Tran, Edeh Michael Onyema, Arij Naser Abougreen

eGovernment Whole-of-Government Approach for Good Governance: The Back-Office Integrated Management IT Systems
Said Azelmad

Advances in Digital Marketing in the Era of Artificial Intelligence: Case Studies and Data Analysis for Business Problem Solving
Moez Ltifi

For more information about this series, please visit: https://www.routledge.com/Advances-in-Digital-Technologies-for-Smart-Applications/book-series/ADT

Data Analytics for Finance Using Python

Nitin Jaglal Untwal and Utku Kose

CRC Press
Taylor & Francis Group
Boca Raton London New York

CRC Press is an imprint of the
Taylor & Francis Group, an **informa** business

First edition published 2025
by CRC Press
2385 NW Executive Center Drive, Suite 320, Boca Raton FL 33431

and by CRC Press
4 Park Square, Milton Park, Abingdon, Oxon, OX14 4RN

CRC Press is an imprint of Taylor & Francis Group, LLC

ISBN: 978-1-032-61821-0 (hbk)
ISBN: 978-1-032-61823-4 (pbk)
ISBN: 978-1-032-61824-1 (ebk)

DOI: 10.1201/9781032618241

Typeset in Caslon
by Apex CoVantage, LLC

Contents

Preface

In today's fast-paced and rapidly changing financial landscape, data analytics has become an essential tool for making informed investment decisions and driving business success. With the increasing availability of financial data, professionals in the finance industry are turning to data analytics to gain a competitive edge and stay ahead of the curve.

This book is designed to provide finance professionals, researchers, and students with a comprehensive guide to the application of data analytics in finance. The book covers a wide range of topics, from the basics of data preprocessing and visualization to advanced machine learning models and inferential statistical techniques for stock price prediction.

This book provides a strong basic foundation for machine learning and its application in finance. It emphasizes various machine learning algorithms and their application to the finance discipline. The advanced machine learning concepts are discussed lucidly over their practical application and theoretical understanding. The topics covered in the book act as a step-by-step guide for the application of machine learning tools in finance. The advanced machine learning topics which are covered are as follows:

Chapter 1: Stock investments portfolio management by applying K-means clustering
Chapter 2: Predicting stock price using the ARIMA model

This book highlights the use and application of machine learning tools and techniques in the finance area to further improve the performance of researchers and analysts. The themes will be helpful for students, researchers, and many other users of these technologies. This book is a catalyst for bringing together machine learning and the financial discipline to develop a deep understanding of the subject.

In addition, we offer our collective expertise and knowledge in this book, providing readers a unique and insightful perspective on finance and computer science and engineering.

Authors

Nitin Jaglal Untwal, PhD, is a distinguished scholar and educator in the field of finance, with a remarkable academic background and research expertise. Holding a doctorate in finance and master's degrees in related fields like commerce, management, and econometrics, he has established himself as a prominent authority in financial data analytics, technology management, and econometrics modeling. With over 11 years of experience in teaching and research, Dr. Untwal has published numerous papers in esteemed databases like Scopus and Web of Science, solidifying his reputation as a leading researcher in his field. Recognized as a postgraduate faculty member by the S.P. University of Pune since 2008, he has also achieved success in prestigious eligibility tests, including UGC-SET in Management and State Eligibility Test Commerce. Additionally, he has completed a Faculty Development Program from the Indian Institute of Management, Kozhikode (IIM-K). Dr. Untwal's wealth of knowledge and experience make him an invaluable contributor to this book.

Utku Kose, PhD, a distinguished scholar in computer science and engineering, joins Dr. Untwal in this literary endeavor. With over 200 publications to his name, Dr. Kose has demonstrated his expertise in artificial intelligence, machine ethics, biomedical applications, and more. His impressive academic background and extensive research experience make him a significant contributor to this book.

1

Stock Investments Portfolio Management by Applying K-Means Clustering

1.1 Introduction

National Stock Exchange Indices is the owner of the Nifty 50 and earlier it was known as Index Services and Product Limited. Nifty 50 covers 12 sectors of the Indian economy. The Nifty 50 is a portfolio of companies from the financial industry, information technology (IT), oil and gas, consumer goods, and automobiles. The composition includes the financial sector with 36.81 percent share, information technology (IT) companies with 14.70 percent share, 12.17 percent share for oil and gas, 9.02 percent share for consumer goods, and 5.84 percent share for automobiles. These companies are considered to be the top performers. Clustering is a technique that classifies the data sets into different groups based on their similarities. The technique of clustering is based on pattern recognition. The Nifty 50 is a group of top-performing companies listed on the National Stock Exchange. The researcher applied K-mean clustering to Nifty 50 stocks to create clusters considering different parameters related to stock valuation. The study is conducted by considering seven parameters: last traded price, price-to-earnings ratio (P/E), debt-to-equity ratio, earning per share (EPS), dividend per share (DPS), return on equity (ROE), and face value. The clustering of high-performing companies is very useful for getting insight into high-value stocks for investors.

The last traded price is the price of a share which is stated at the end of the day. It is the price that occurred as the last traded price. It differs from the closing price. Price-to-earnings ratio is the ratio of the current market price of the share to earnings per share, also called price multiple ratio. It is a handy tool for comparing the price and performance of different stocks. It majors the proportion of a

DOI: 10.1201/9781032618241-1

company's stock price to earnings per share. A high P/E ratio indicates that a company's stock is overvalued or that the investors expect high growth rates.

Debt-to-equity ratio is the ratio of total debt to total shareholders' equity. It is the ratio of borrowed capital to the owned capital. Higher debt-to-equity ratio means that a company is using borrowed capital for financing. The debt-to-equity ratio of 1 to 1.5 is considered to be the standard ratio. The depth-to-equity ratio may vary from industry to industry.

Earnings per share are calculated by dividing the earnings before interest and tax by the number of shareholders. The company's financial position is reflected in its EPS. If the EPS is high, it means that shareholders' value is increased, which is considered to be the main objective of financial management.

Dividend per share in which dividend is the reward to shareholders for investing and taking risks. It is the dividend issued divided by the number of shareholders. The dividend per share is based on the amount of dividend issued from the overall earnings of the company. The retained earnings are kept aside keeping the future growth and expansion plan of the company, further, it plays an important role in deciding the Dividend policy.

Return on equity is the percentage of net income to the value of shareholders' equity. Higher the percentage, more efficient is the generation of profit.

1.1.1 Introduction to Cluster Analysis

Machine learning: The unsupervised learning models are trained for data sets, which are unlabeled and are allowed to act without any supervision. The grouping of data sets into different clusters helps to generate different meaningful patterns to analyze unlabeled data sets. The clustering machine learning model is an unsupervised learning model that helps in determining the patterns of unlabeled data sets. To find out similar patterns or dissimilar patterns in a data set clustering technique is applied. Every data set has some common features based on which we can draw similar patterns, categories and group the data into different clusters. The K-means technique is one of the very popular clustering techniques for analysis. This is the reason for its highest usage and application. The K-means clustering is very simple to understand and apply. It is one of the best partitioning techniques for data analysis. The cluster technique is based on the concept of centroid which makes clustering formation unique.

The K-means clustering is the technique of grouping and classifying the data sets into different categories based on the nearest distance from the mean. The clustering produces the exact number of clusters of the greatest possible difference, which is known as priori. It also controls the total cluster and the total cluster variance.

It is represented by the equation

$$j = \sum_{j=1}^{k} \sum_{i=1}^{n} \left\| x_i^{(j)} - c_j \right\|^2$$

Here, J is the objective function, k stands for cluster numbers, n denotes the number of cases, x is the number of cases i, and c_j is the number of the centroid.

1.1.2 Literature Review

The cluster technique has been applied to study the financial market. Bonanno et al. (2003) worked on network structures of equities and found out the relationship between them further by applying a complex system. Coelho et al. (1996) analyzed large noisy data sets by applying cluster analysis. Jain (2010) and Nanda (2010) applied cluster analysis for portfolio management (Coronnello et al., 2005). Madhavan (2000) applied clustering to study the market microstructure. Onnela et al. (2003) and Bonanno et al. (2001) explored the correlation between different markets (Huang et al., 2011). Mantegna (1999) studied portfolio management strategies for financial forecasting and analysis. Song et al. (2011) applied random matrix theory to develop insights into the movement of the financial market (Kantar & Deviren, 2014; Kenett et al., 2011).

1.2 Research Methodology

1.2.1 Data Source

Nifty 50 database

1.2.2 Study Time Frame

The data selected for analysis is the ratio of different companies under Nifty 50.

1.2.3 Tool for Analysis

Python Programming

1.2.4 Model Applied

For this study, we applied K-means clustering.

1.2.5 Limitations of the Study

The study is restricted to cluster analysis for only Nifty 50 companies.

1.2.6 Future Scope

A similar kind of cluster analysis can be done for the different sectors of the Indian economy at the macro level.

Research Is Carried Out in Five Steps

1.3 Feature Extraction and Engineering
1.4 Data Extraction
1.5 Standardizing and Scaling
1.6 Identification of Clusters
1.7 Cluster Formation

1.3 Feature Extraction and Engineering

Feature engineering is an important element of the machine learning pipeline. Python directly cannot read a particular file as it does not suit the Python environment; hence, we need to fetch the data into the Python environment. Feature engineering is a process that creates data that can be utilized by the machine learning algorithm for analysis. The raw data cannot be used as they contain so many errors. The raw data need to be transformed depending upon the domain knowledge and also according to the need for machine learning modeling. We cannot create a good machine learning model unless we have done feature engineering. Feature engineering is the process of cleaning, and defining the data into different parameters to make it algorithm-friendly.

1.4 Data Extraction

The process of fetching data from an external source into the Python environment and further making it readable by the Jupiter environment to carry out machine learning analysis is known as data extraction (Refer Figure 1.1). For this study, we need to fetch the Excel file which contains the financial information about the Nifty 50 companies. We use different libraries for K-means clustering in Python such as Pandas, matplotlib, and sklearn.

```
In [1]:  # importing required libraries
         import pandas as pd
         import numpy as np
         import matplotlib.pyplot as plt
         %matplotlib inline
         from sklearn.cluster import KMeans
```

Figure 1.1   Data fetching in the Python environment.

Data is cleaned by removing the unwanted column from the data frame by applying the Python code below (Refer Figure 1.2):

```
In [2]:  import pandas as pd

         data = pd.read_excel (r'C:\Users\nitin\Desktop\niftycls.xlsx')
         print (data)
```

	LTP	P/E	Debt to Equity	EPS(Rs.)	DPS(Rs.)	ROE%	\
0	2930.80	307.60	0.81	6.55	1.00	13.75	
1	1071.25	549.08	1.68	1.41	5.00	1.11	
2	5749.95	97.65	0.33	46.25	11.75	10.88	
3	3391.25	94.25	0.00	32.68	19.15	23.48	
4	1091.90	17.92	0.00	42.48	1.00	11.32	
5	6675.00	21.04	0.00	173.60	140.00	18.81	
6	7407.50	78.21	2.78	65.85	10.00	11.00	
7	1685.80	863.27	0.00	11.20	3.00	4.70	
8	1024.95	-115.61	1.31	-6.53	3.00	-4.59	
9	456.45	8.70	0.49	41.31	16.00	17.69	
10	5287.05	48.17	0.91	66.56	56.50	66.72	
11	1282.20	27.80	0.00	36.62	0.00	13.13	
12	393.80	10.07	0.00	18.18	17.00	68.47	
13	4014.65	39.63	0.00	111.07	30.00	25.21	
14	5944.40	43.90	0.12	97.85	30.00	8.85	
15	3897.55	42.35	0.00	58.02	21.00	14.69	
16	2109.45	35.81	0.08	46.47	10.00	6.27	
17	1473.60	29.02	0.01	40.10	42.00	25.53	
18	1696.80	22.01	0.00	66.80	15.50	15.39	
19	645.40	103.45	0.05	6.73	2.02	12.15	
20	4115.15	18.53	0.00	123.78	95.00	15.66	
21	618.60	16.38	0.35	34.76	4.00	10.11	
22	2615.95	54.59	0.00	37.53	34.00	18.08	
23	985.25	21.70	0.00	33.66	5.00	13.68	
24	1585.80	15.70	0.00	59.57	8.50	9.66	
25	1533.95	37.77	0.00	50.49	31.00	30.63	
26	470.85	20.51	0.00	12.22	11.50	24.52	
27	870.50	10.54	0.79	69.48	17.35	26.30	
28	1874.30	49.84	0.00	35.17	0.90	11.01	

Figure 1.2   Data frame for Nifty 50 in the Python environment.

1.5 Standardizing and Scaling

Before performing a machine learning algorithm on a particular data set, we need to standardize and scale the data as a huge amount of variation is present in the data set. The huge amount of variation needs to be transformed by scaling to remove the difference in the magnitude of the data set (Refer Figure 1.3). Furthermore, this will remove the difficulty of variation as the K-means clustering algorithm is distance based (Pozzi et al., 2012). Scaling is applied since we need to standardize the data by bringing down the standard deviation and mean of features to one and zero (Lillo & Mantegna, 2003; Peralta & Zareei, 2016).

Python code is applied as follows:

```
In [3]:  # statistics of the data
         data.describe()
```

Out[3]:

	LTP	P/E	Debt to Equity	EPS(Rs.)	DPS(Rs.)	ROE%	Face Value
count	50.000000	50.000000	50.000000	50.000000	50.000000	50.000000	50.00000
mean	3008.194000	15.230000	0.319200	58.277000	25.122400	18.744600	4.22000
std	4270.271041	363.546356	0.580914	61.287728	36.269028	18.548535	3.57594
min	139.900000	-2286.880000	0.000000	-6.530000	0.000000	-6.970000	1.00000
25%	807.350000	14.357500	0.000000	19.185000	6.012500	10.910000	1.00000
50%	1559.875000	28.410000	0.020000	40.690000	11.525000	14.040000	2.00000
75%	3769.550000	49.422500	0.322500	66.382500	30.750000	23.047500	5.00000
max	27240.000000	863.270000	2.780000	270.330000	200.000000	102.890000	10.00000

```
In [4]:  # standardizing the data
         from sklearn.preprocessing import StandardScaler
         scaler = StandardScaler()
         data_scaled = scaler.fit_transform(data)

         # statistics of scaled data
         pd.DataFrame(data_scaled).describe()
```

Out[4]:

	0	1	2	3	4	5	6
count	5.000000e+01	5.000000e+01	5.000000e+01	5.000000e+01	5.000000e+01	5.000000e+01	5.000000e+01
mean	1.154632e-16	-5.162537e-17	4.107825e-17	-1.564312e-17	3.108624e-17	-1.110223e-16	1.243450e-16
std	1.010153e+00	1.010153e+00	1.010153e+00	1.010153e+00	1.010153e+00	1.010153e+00	1.010153e+00
min	-6.785083e-01	-6.396660e+00	-5.550579e-01	-1.068158e+00	-6.997005e-01	-1.400416e+00	-9.096045e-01
25%	-5.206199e-01	-2.424335e-03	-5.550579e-01	-6.443196e-01	-5.322424e-01	-4.266720e-01	-9.096045e-01
50%	-3.426066e-01	3.662204e-02	-5.202799e-01	-2.898713e-01	-3.787101e-01	-2.562124e-01	-6.271186e-01
75%	1.801023e-01	9.500753e-02	5.738381e-03	1.335959e-01	1.567380e-01	2.343358e-01	2.203390e-01
max	5.732147e+00	2.356370e+00	4.279093e+00	3.495086e+00	4.870631e+00	4.582556e+00	1.632768e+00

Figure 1.3   Descriptive statistics for scaled data.

1.6 Identification of Clusters by the Elbow Method

The elbow technique is applied in Python Programming to get a defined number of clusters. The elbow technique is a diagrammatic representation of cluster formation from a given data set. It works on the concept of centroid (Song et al., 2011). It explains the variation in different data sets and finds out the exact number of clusters. The clusters explains variation which also exempts overfitting. It also removes the various constraints in the cluster. Identification and its formation. It also optimizes the number of clusters by applying inertia which acts as a tool for well-defined clusters by applying clustering technique. The elbow technique optimizes the number of clusters with lower inertia and less number of clusters (Peralta & Zareei, 2016).

The calculation of within-cluster inertia is

$$Inertia\ (k) = \sum_{i\ \backslash in\ C\_k}\ (y\_{ik}-\mu\_k)\ ^{\wedge}2$$

where μ_k is the mean of cluster k and C_k corresponds to the set of indices of genes attributed to cluster k.

1.7 Cluster Formation

When we apply the Python code shown below (Refer Figure 1.4), we get the results for clusters one to six according to their characteristics and features.

In [7]:
```
# fitting multiple k-means algorithms and storing the values in an empty list
SSE = []
for cluster in range(1,20):
    kmeans = KMeans(n_jobs = -1, n_clusters = 6 , init='k-means++')
    kmeans.fit(data_scaled)
    SSE.append(kmeans.inertia_)

# converting the results into a dataframe and plotting them
frame = pd.DataFrame({'Cluster':range(1,20), 'SSE':SSE})
plt.figure(figsize=(12,6))
plt.plot(frame['Cluster'], frame['SSE'], marker='o')
plt.xlabel('Number of clusters')
plt.ylabel('Inertia')
```
Out[7]: Text(0,0.5,'Inertia')

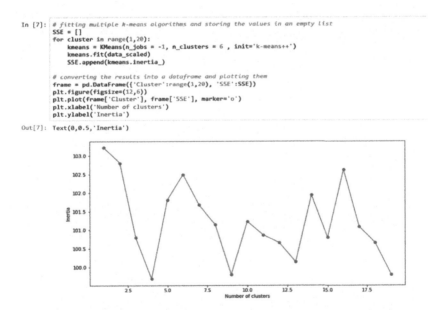

Figure 1.4   Cluster formation Elbow technique.

8 DATA ANALYTICS FOR FINANCE USING PYTHON

```
In [8]:  # k means using 5 clusters and k-means++ initialization
         kmeans = KMeans(n_jobs = -1, n_clusters = 6, init='k-means++')
         kmeans.fit(data_scaled)
         pred = kmeans.predict(data_scaled)
```

```
In [9]:  frame = pd.DataFrame(data_scaled)
         frame['cluster'] = pred
         frame['cluster'].value_counts()
```

```
Out[9]:  1    26
         0    10
         5     6
         4     6
         3     1
         2     1
         Name: cluster, dtype: int64
```

Figure 1.5   Cluster formation results with classification for scaled data.

1.8 Results and Analysis

We categorized the data selected for analysis into six clusters in Python (Refer Figure 1.5 and Figure 1.6):

Figure 1.6   Cluster analysis.

1.8.1 Cluster One

After applying clustering, the result shows that there are six clusters. Cluster one includes companies like Bajaj Auto, Britannia, Divis Laboratories, Dr. Reddy's lab, Hero Motors, lit Maitri, Maruti Suzuki, Tata Steel, TCS, and UltraTech Cement (Refer Table 1.1). The average LTE for cluster one is 5653, the maximum LTE is 10,209, and the minimum LTE is 139. Cluster one includes companies from the automobile sector, cement sector, and pharma, and only one company from food processing sector, which is Britannia.

The price-to-earnings ratio of Maruti Suzuki is the highest at 60 percent. The lowest price-to-earnings ratio is registered with Tata Steel with a value of 4.84. The debt-to-equity ratio of almost all companies is less than 0 only Britannia had a registered debt-to-equity ratio of 0.91. The highest earning per share is registered with Tata Steel with a value of 270. The minimum Earning per share is registered with the value of 66.56 for Britannia. The dividend per share of 140 is registered with Bajaj Auto which is the highest dividend after Hero Motors. Cluster one the automobile companies like Bajaj Auto, Hero Motors, and Maruti Suzuki had given a good dividend. Pharma companies in cluster one had a good LTP but their dividend payout ratio was considerably low in comparison to Pharma. In cluster one, Britannia dominated all financial parameters and showed a sound financial position.

Table 1.1   Cluster One Classifications for Nifty 50 Companies

NAME	LTP	P/E	DEBT TO EQUITY	EPS (RS.)	DPS (RS.)	ROE %	FACE VALUE	CLUSTER
Bajaj Auto	6,675.00	21.04	0	173.6	140	18.81	10	0
Britannia	5,287.05	48.17	0.91	66.56	56.5	66.72	1	0
Divis Labs	4,014.65	39.63	0	111.07	30	25.21	2	0
Dr Reddys Labs	5,944.40	43.9	0.12	97.85	30	8.85	5	0
Hero Motocorp	4,115.15	18.53	0	123.78	95	15.66	2	0
LTIMindtree	6,153.05	47.66	0	129.14	55	26.9	1	0
Maruti Suzuki	10,209.50	60.65	0.01	124.68	60	6.96	5	0
Tata Steel	139.9	4.84	0.29	270.33	51	26.31	10	0
TCS	3,790.40	35.84	0	104.34	43	49.48	1	0
UltraTechCement	10,205.00	26.95	0.2	245	38	14.34	10	0
Avg	5,653.41	34.721	0.153	144.635	59.85	25.924	4.7	
Max	10,209.50	60.65	0.91	270.33	140	66.72	10	
Min	139.90	4.84	0	66.56	30	6.96	1	

1.8.2 Cluster Two

In cluster 2 the average LTE is 1822 with the maximum of 5749 for Apollo hospital. In cluster two the service sector bank which includes Axis Bank, HDFC Bank, SBI, ICICI Bank had been the key players in cluster 2 represented by the banking sector (Refer Table 1.2). The Apollo Hospital has a good LTP ratio, Debt to the equity ratio and Earnings per share is leading Cluster Two with all financial parameters. Cluster two also includes information technology companies like Wipro, Tech Mahindra, Infosys which are below Apollo Hospital which is a service sector company.

Table 1.2　Cluster Two Classifications for Nifty 50 Companies

NAME	LTP	P/E	DEBT TO EQUITY	EPS (RS.)	DPS (RS.)	ROE %	F.V
Adani Enterprise	2,930	307.6	0.81	6.55	1	13.75	1
Apollo Hospital	5,749	97.65	0.33	46.25	11.75	10.88	5
Asian Paints	3,391	94.25	0	32.68	19.15	23.48	1
Axis Bank	1,091	17.92	0	42.48	1	11.32	2
Bajaj Finserv	1,685	863.27	0	11.2	3	4.7	5
Cipla	1,282	27.8	0	36.62	0	13.13	2
Eicher Motors	3,897	42.35	0	58.02	21	14.69	1
Grasim	2,109	35.81	0.08	46.47	10	6.27	2
HCL Tech	1,473	29.02	0.01	40.1	42	25.53	2
HDFC Bank	1,696	22.01	0	66.8	15.5	15.39	1
Hindalco	618	16.38	0.35	34.76	4	10.11	1
HUL	2,615	54.59	0	37.53	34	18.08	1
ICICI Bank	985	21.7	0	33.66	5	13.68	2
Infosys	1,533	37.77	0	50.49	31	30.63	5
ETC	470	20.51	0	12.22	11.5	24.52	1
JSW Steel	870	10.54	0.79	69.48	17.35	26.3	1
Kotak Mahindra	1,874	49.84	0	35.17	0.9	11.01	5
Larsen	3,447	31.51	0.3	56.09	22	11.74	2
M&M	1,656	19.54	0.17	41.28	11.55	12.66	5
ONGC	208	5.12	0.03	32.04	10.5	16.99	5
SBI	642.8	13.91	0	35.49	7.1	11.3	1
TATA Cons. Prod	1,098	80.89	0	9.61	6.05	7.53	1
Tech Mahindra	1,279	29.7	0	50.48	45	19	5
Titan Company	3,707	103.26	0.02	24.56	7.5	23.25	1
UPL	595	50.01	0.2	15.39	10	14.33	2
Wipro	470	26.66	0.14	22.2	6	22.32	2

Table 1.2 (Continued)

NAME	LTP	P/E	DEBT TO EQUITY	EPS (RS.)	DPS (RS.)	ROE %	F.V
Avg	1,822	81.13	0.12	36.44	13.60	15.86	2.38
Max	5,749	863	0.81	69.48	45	30.63	5
Min	208.07	5.12	0	6.55	0	4.7	1

1.8.3 Clusters Three and Four

Clusters three and four include companies like Sun Pharma and Nestle. Further, the EPS for Sun Pharma is negative and EPS for Nestle is 222 with a dividend per share of 200 (Refer Table 1.3).

Table 1.3   Clusters Three and Four Classifications for Nifty 50 Companies

NAME	LTP	P/E	DEBT TO EQUITY	EPS (RS.)	DPS (RS.)	ROE %	FACE VALUE	CLUSTER
Sun Pharma	1,298.80	−2,286.88	0.2	−0.4	10	−0.4	1	3
Nestle	27,240.00	0	0.02	222.46	200	102.89	10	4

1.8.4 Cluster Five

In cluster Five the highest LTP is registered with Bajaj Finance with 7407. The P/E ratio for Bajaj Finance is 78 and the very high debt-to-equity ratio of 2.78 (Refer Table 1.4). The EPS for Bajaj Finance is highest in cluster Five with 65. The maximum return on equity registered with 22.4 by Power Grid Corporation which is highest in cluster Five

1.8.5 Cluster Six

In Cluster six which includes Three Service sector banks and insurance companies (Refer Table 1.5). The highest LTP is registered with Reliance with a value of 20601 and the lowest last credit price is registered for the public sector organization which is Coal India. The highest price-to-earnings ratio is registered with HDFC Life Insurance. The highest debt-to-equity ratio is registered with BPCL. For the highest earning per share is registered with the Indus bank. The

Table 1.4 Cluster Five Classifications for Nifty 50 Companies

NAME	LTP	P/E	DEBT TO EQUITY	EPS (RS.)	DPS (RS.)	ROE %	FACE VALUE	CLUSTER
Adani Ports	1,071.25	549.08	1.68	1.41	5	1.11	2	5
Bajaj Finance	7,407.50	78.21	2.78	65.85	10	11	2	5
Bharti Airtel	1,024.95	−115.61	1.31	−6.53	3	−4.59	5	5
NTPC	306.65	8.12	1.33	16.62	7	12.58	10	5
Power Grid Corp	239	8.85	1.77	24.51	14.75	22.44	10	5
Tata Motors	786.3	−119.49	1.16	−3.63	0	−6.97	2	5
Avg	1,805.94	68.1933333	1.671666667	16.37167	6.625	5.928333	5.166666667	
Max	7,407.50	549.08	2.78	65.85	14.75	22.44	10	
Min	239.00	−119.49	1.16	−6.53	0	−6.97	2	

Table 1.5 Cluster Six Classifications for Nifty 50 Companies

NAME	LTP	P/E	DEBT TO EQUITY	EPS (RS.)	DPS (RS.)	ROE %	FACE VALUE	CLUSTER
BPCL	456.45	8.7	0.49	41.31	16	17.69	10	6
Coal India	393.8	10.07	0	18.18	17	68.47	10	6
HDFC Life	645.4	103.45	0.05	6.73	2.02	12.15	10	6
IndusInd Bank	1,585.80	15.7	0	59.57	8.5	9.66	10	6
Reliance	2,601.15	44.48	0.41	59.24	8	8.28	10	6
SBI Life Insurance	1,434.30	0	0	14.56	2.5	11.09	10	6
Avg	1186.15	30.4	0.158333333	33.265	9.003333	21.22333	10	
Max	2601.15	103.45	0.49	59.57	17	68.47	10	
Min	393.8	0	0	6.73	2.02	8.28	10	

highest dividend per share is registered with Coal India the highest return on equity is also given by Coal India with 68.47.

1.9 Conclusion

The overall cluster formation is classified into six clusters based on different parameters like Last Traded Price, Price earnings ratio

(P/E), Debt to Equity, Earning per share (EPS), Dividend per share (DPS), return on equity (ROE), and Face Value. The clustering of high-performing companies is very useful for getting insight into high-value stocks for investors.

References

Bonanno, G., Caldarelli, G., Lillo, F., Miccichè, S., Vandewalle, N., & Mantegna, R. N. (2003). Networks of equities in financial markets. The European Physical Journal B-Condensed Matter and Complex Systems, 38(2), 363–371.

Bonanno, G., Lillo, F., & Mantegna, R. N. (2001). High-frequency cross-correlation in a set of stocks. Quantitative Finance, 1(1), 96–104.

Coelho, R., Gilmore, C. G., Lucey, B. M., Richmond, P., & Hutzler, S. (2007). The evolution of interdependence in world equity markets—Evidence from minimum spanning trees. Physica A: Statistical Mechanics and its Applications, 376, 455–466.

Coronnello, C., Tumminello, M., Lillo, F., Miccichè, S., & Mantegna, R. N. (2005). Sector identification in a set of stock return time series: A comparative study. Quantitative Finance, 5(4), 373–387.

Ester, M., Kriegel, H. P., Sander, J., & Xu, X. (1996). A density-based algorithm for discovering clusters in large spatial databases with noise. In Kdd (Vol. 96, No. 34, pp. 226–231).

Huang, Z., Cai, Y., & Xu, X. (2011). A data mining framework for investment opportunities identification. KDD-96: The Second International Conference on Knowledge Discovery and Data Mining. Expert Systems with Applications, 38(8), 9224–9233.

Jain, A. K. (2010). Data clustering: 50 years beyond K-means. Pattern Recognition Letters, 31(8), 651–666.

Kantar, E., & Deviren, B. (2014). Hierarchical structure of stock markets. Physica A: Statistical Mechanics and Its Applications, 404, 117–128.

Kenett, D. Y., Shapira, Y., & Ben-Jacob, E. (2011). RMT assessments of the market latent information embedded in the stocks' raw data. Journal of Probability and Statistics. DOI:10.1155/2009/249370 (2009)

Lillo, F., & Mantegna, R. N. (2003). Power-law relaxation in a complex system: Omori law after a financial market crash. Physical Review E, 68(1), 016119.

Madhavan, A. (2000). Market microstructure: A survey. Journal of Financial Markets, 3(3), 205–258.

Mantegna, R. N. (1999). Hierarchical structure in financial markets. The European Physical Journal B, 11(1), 193–197.

Nanda, S., Mahanty, B., & Tiwari, M. K. (2010). Clustering Indian stock market data for portfolio management. Expert Systems with Applications, 37(12), 8793–8798.

Onnela, J. P., Chakraborti, A., Kaski, K., Kertész, J., & Kanto, A. (2003). Dynamics of market correlations: Taxonomy and portfolio analysis. Physical Review E, 68(5), 056110.

Peralta, G., & Zareei, A. (2016). A network approach to portfolio selection. Journal of Empirical Finance, 38, 157–180.

Pozzi, F., Di Matteo, T., & Aste, T. (2012). Exponential smoothing weighted correlations. The European Physical Journal B, 85(6), 175.

Song, D. M., Tumminello, M., Zhou, W. X., & Mantegna, R. N. (2011). Evolution of worldwide stock markets, correlation structure, and correlation-based graphs. Physical Review E, 84(2), 026108.

2

PREDICTING STOCK PRICE USING THE ARIMA MODEL

2.1 Introduction

The stock is exposed to different types of risk and uncertainties which have an impact on the price of the stock. It is difficult to predict the stock price. The stock price is influenced by various factors related to demand and supply. For predicting the price of a stock, we require dependent variables like stock market index, similar or identical company, sales, profits, earnings per share, etc. When the dependent variable does not have any impact and it is impossible to predict the stock price in such cases, the stock price is predicted by considering the past stock value on different time horizons like days, week, months, quarterly, or yearly by applying the autoregressive moving average method called ARIMA.

Stock prediction using time series analysis is an emerging area in predictive analytics. It has attracted many researchers because of the utility and accuracy of the model. The main objective of the ARIMA model is to study past observations based on which future models are generated to forecast a given variable. The success of the ARIMA model depends on appropriate model identification and evaluation.

It is essential to understand that the ARIMA model is applied under what circumstances.

- No dependent variable is available.
- Good sufficient historical data is available.
- Autocorrelation.

2.2 ARIMA Model

The ARIMA model has a wide area of application for estimating and predicting the future value of a variable in applied econometrics areas like management, finance, banking, health analytics, and weather forecasting

DOI: 10.1201/9781032618241-2

which are crucial for selecting the optimized model that can predict the precise value of a given variable depending on historical and past data. The ARIMA model is considered to be the most reliable model in such a situation; Box and Jenkins is the researchers and scientists who developed the ARIMA model in 1970. The model was used in forecasting and showed tremendous potential to generate a short-term forecast.

The ARIMA model forecasts time lags which are equally spaced in time horizon with univariate time series. In ARIMA, AR stands for autoregressive, which emphasizes on the relationship between the past values and the future values, I stands for integrated, and MV stands for moving average.

It is represented by the equation

$$yt = \Phi 0 + \Phi 1 \ it-1 + \Phi 2 \ it-2 +. . .+ \Phi pat-p$$
$$+ \epsilon t-\theta 1 \ \epsilon \ t-1-\theta 2 \ \epsilon \ t-2-. . .-\theta q \ \epsilon \ t-q \qquad (2.1)$$

where actual data values are denoted as yt, coefficients are denoted as Φi and θj, ϵi denotes the random errors, and integers p and q represent the degrees of autoregressive and moving averages (Ayodele et al., 2014). The ARIMA model is a mixture of two equations: Autoregressive is the equation based on past lags and the moving average is based on error.

2.2.1 Literature Review

Book explains. Various predictive models, including ARIMA for its practical application and use Burbidge et al. (2001). The time series analysis and its application for forecasting is developed and explained lucidly Burges (1998). The main focus is on forecasting of financial data and risk analysis associated with it Cervantes et al. (2023). Applied hybrid autoregressive integrated moving average method with a neural network model further explains the combination of both the models and has improved the predictive analysts forecasting model Deo (2015). Developed hybrid model includes long-term short-term memory, autoregressive integrated moving average, and the Bayes optimization model for predicting the financial data in the form of forecasting stock price Dhillon and Verma (2020).

Furthermore, various research studies focused on hybrid models of LSTM and machine learning with time series analysis (Ding &

Dubchak, 2001; Drucker et al., 1999; Garcia-Lamont et al., 2023; Hinton et al., 2012; Huang et al., 2005; Joachims, 1998; Kim, 2014; Maita et al., 2015; Mountrakis et al., 2011; Nguyen et al., 2020; Pal & Mather, 2003; Schölkopf et al., 2001; Tay & Cao, 2001; Toledo-Pérez et al., 2019; Turk & Pentland, 1991).

2.3 Research Methodology

2.3.1 Data Source

Yahoo Finance financial database is used to create the ARIMA model.

2.3.2 Period of Study

The study period was from 8 January 2023 to 5 January 2024. The interval for the selected data is the daily closing stock price of MRF for analysis.

2.3.3 Software Used for Data Analysis

Python Programming, Anaconda

2.3.4 Model Applied

For this study, we applied the ARIMA model.

2.3.5 Limitations of the Study

The study is restricted to the Stock price Index of MRF only.

2.3.6 Future Scope of the Study

In the future, the study can be done on the macro level by applying it to a different stock at the same time.

2.3.7 Methodology

The autoregressive integrated moving average (ARIMA) model predicts the stock price by using past values. The ARIMA model is applied to understand the relationship between the past values of stock for predicting its future predicted value. The ARIMA model is widely applied in the field of stock price prediction. The ARIMA model is implemented first by understanding the relationship between the past values of stock and its future value. Autocorrelation plays an important role in model development. The check for autocorrelation defines further steps of model evaluation and parameter estimation to select the best ARIMA model for stock prediction using Python. Research is carried out in three steps. First, we need to check the autocorrelation. Then, we need to evaluate different ARIMA models and compare the AIC of other models. The best model is selected with the lowest AIC and mean square error given by train and test data analysis results. The data set is divided into two parts: 70 percent train data and 30 percent test data.

Research is carried out in three steps:

2.4 Finding Different Lags Autocorrelation
2.5 Creating the Different ARIMA Models
2.6 Selecting the Best Model Using Cross-Validation

2.4 Finding Different Lags Autocorrelation

The autocorrelation at different lags (lag1, lag2, lag3, lag4, lag5, etc.) are considered by using matplot, and autocorrelation is detected by studying the projections in autocorrelation charts. Autocorrelation is the relationship between successive values of the same variable. Here, we will cross-check the autocorrelation in our time series data using Python Programming and comparing the autocorrelation at different lags (Figures 2.1–2.5 show the autocorrelation plot for lag = 1 to lag = 5 for the MRF stock).

The plot in Figure 2.1 shows a very high degree of autocorrelation for lag = 1; hence, we further checked the autocorrelation for lag = 2.

The plot in Figure 2.2 does not show the substantial degree of autocorrelation for lag = 2; hence, we further checked the autocorrelation for lag = 3.

The plot in Figure 2.3 does not show a considerable degree of autocorrelation for lag = 3; hence, we further checked the autocorrelation for lag = 4.

```
In [3]: plt.figure()
        lag_plot(DEF_data['price'], lag=1)
        plt.title('DEF - Autocorrelation plot with lag = 1')
        plt.show()
```

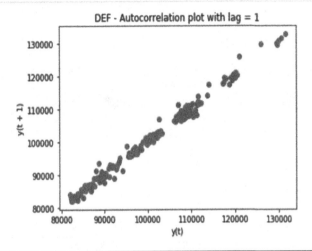

Figure 2.1   An autocorrelation plot with lag = 1 for the MRF stock.

```
In [4]: plt.figure()
        lag_plot(DEF_data['price'], lag=2)
        plt.title('DEF - Autocorrelation plot with lag = 2')
        plt.show()
```

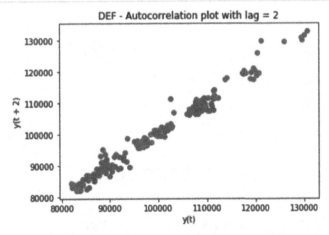

Figure 2.2   An autocorrelation plot with lag = 2 for the MRF stock.

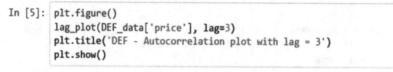

```
In [5]: plt.figure()
        lag_plot(DEF_data['price'], lag=3)
        plt.title('DEF - Autocorrelation plot with lag = 3')
        plt.show()
```

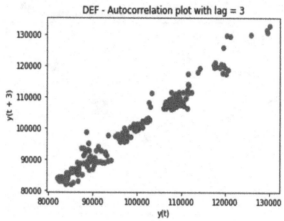

Figure 2.3 An autocorrelation plot with lag = 3 for the MRF stock.

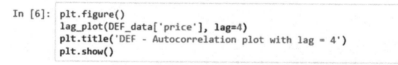

```
In [6]: plt.figure()
        lag_plot(DEF_data['price'], lag=4)
        plt.title('DEF - Autocorrelation plot with lag = 4')
        plt.show()
```

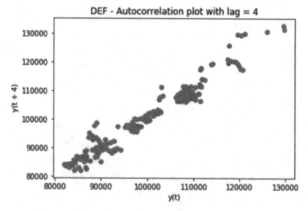

Figure 2.4 An autocorrelation plot with lag = 4 for the MRF stock.

The plot in Figure 2.4 shows a lesser degree of autocorrelation for lag = 4; hence, we further checked the autocorrelation for lag = 5.

The plot in Figure 2.5 does not show a substantial degree of auto-correlation for lag = 5. The above analysis from lag1 to lag5 shows that

```
In [7]:  plt.figure()
         lag_plot(DEF_data['price'], lag=5)
         plt.title('DEF - Autocorrelation plot with lag = 1')
         plt.show()
```

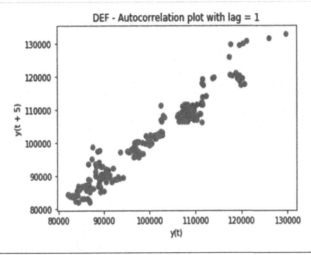

Figure 2.5   An autocorrelation plot with lag = 5 for the MRF Stock.

lag1 has the highest autocorrelation (Figures 2.1–2.5 show the auto-correlation plot for lag = 1 to lag = 5 for the MRF stock).

Akaike information criterion (AIC) for ARIMA model evaluation is considered to be the best method and hence we tried to compare the AIC of different ARIMA models.

2.5 Creating the Different ARIMA Models

We first find out the Akaike Information Criterion (AIC) for model evaluation. The ARIMA models are compared to check the AIC to select the best model. It determines the order of an ARIMA model.

AIC is given by the following equation:

$$AIC = -2 \, Log \, (L) + 2 \, (p+q+k+1). \ldots \ldots \qquad (2.2)$$

where L is the likelihood of the data; k=1 if c ≠ 0; and k = 0 if c = 0. We tested four ARIMA models, and Figures 2.6–2.10 show the details of different ARIMA models as Python Programming output. The ARIMA (1,1,1) model has an AIC value of 7459.

Inference—The ARIMA (1,1,1) model has an AIC value of 7459 (Refer Figure 2.6).

```
In [10]: from statsmodels.tsa.arima_model import ARIMA
         # 1,1,1 ARIMA Model
         model = ARIMA(history, order=(1,1,1))
         model_fit = model.fit(disp=0)
         print(model_fit.summary())
```

```
                            ARIMA Model Results
==============================================================================
Dep. Variable:                   D.y   No. Observations:                  424
Model:                 ARIMA(1, 1, 1)   Log Likelihood               -3725.911
Method:                       css-mle   S.D. of innovations           1585.390
Date:                Mon, 08 Jan 2024   AIC                           7459.822
Time:                        14:24:46   BIC                           7476.020
Sample:                             1   HQIC                          7466.222

==============================================================================
                 coef    std err          z      P>|z|      [0.025      0.975]
------------------------------------------------------------------------------
const         105.1971     76.592      1.373      0.170     -44.920     255.314
ar.L1.D.y      -0.8279      0.405     -2.043      0.042      -1.622      -0.034
ma.L1.D.y       0.8184      0.415      1.974      0.049       0.006       1.631
                                    Roots
==============================================================================
                  Real          Imaginary           Modulus         Frequency
------------------------------------------------------------------------------
AR.1           -1.2078           +0.0000j            1.2078            0.5000
MA.1           -1.2219           +0.0000j            1.2219            0.5000
------------------------------------------------------------------------------
```

Figure 2.6   Results for the ARIMA (1,1,1) model.

Inference—The ARIMA (1,0,2) model has an AIC value of 7485 (Refer Figure 2.7).

```
In [12]: from statsmodels.tsa.arima_model import ARIMA
         # 1,0,2 ARIMA Model
         model = ARIMA(history, order=(1,0,2))
         model_fit = model.fit(disp=0)
         print(model_fit.summary())
```

```
                            ARMA Model Results
==============================================================================
Dep. Variable:                     y   No. Observations:                  425
Model:                    ARMA(1, 2)   Log Likelihood               -3737.829
Method:                      css-mle   S.D. of innovations           1588.210
Date:               Mon, 08 Jan 2024   AIC                           7485.658
Time:                       14:26:13   BIC                           7505.919
Sample:                            0   HQIC                          7493.662

==============================================================================
                 coef    std err          z      P>|z|      [0.025      0.975]
------------------------------------------------------------------------------
const       9.855e+04   1.45e+04      6.798      0.000    7.01e+04    1.27e+05
ar.L1.y        0.9955      0.006    169.653      0.000       0.984       1.007
ma.L1.y       -0.0041      0.049     -0.084      0.933      -0.100       0.092
ma.L2.y        0.0212      0.052      0.406      0.685      -0.081       0.124
                                    Roots
==============================================================================
                  Real          Imaginary           Modulus         Frequency
------------------------------------------------------------------------------
AR.1            1.0045           +0.0000j            1.0045            0.0000
MA.1            0.0971           -6.8671j            6.8678           -0.2477
MA.2            0.0971           +6.8671j            6.8678            0.2477
------------------------------------------------------------------------------
```

Figure 2.7   Results for the ARIMA (1,0,2) model.

Inference—The ARIMA (0,0,3) model has an AIC value of 8058 (Refer Figure 2.8).

```
In [16]:  from statsmodels.tsa.arima_model import ARIMA
          # 0,0,3 ARIMA Model
          model = ARIMA(history, order=(0,0,3))
          model_fit = model.fit(disp=0)
          print(model_fit.summary())
```

```
                         ARMA Model Results
============================================================================
Dep. Variable:                    y   No. Observations:              425
Model:                    ARMA(0, 3)   Log Likelihood           -4024.226
Method:                     css-mle   S.D. of innovations        3122.158
Date:             Mon, 08 Jan 2024   AIC                        8058.452
Time:                      14:28:46   BIC                        8078.713
Sample:                           0   HQIC                       8066.456

============================================================================
                 coef    std err          z      P>|z|      [0.025      0.975]
----------------------------------------------------------------------------
const         9.862e+04    662.141    148.939      0.000    9.73e+04    9.99e+04
ma.L1.y          1.4754      0.043     34.369      0.000       1.391       1.560
ma.L2.y          1.2982      0.043     30.082      0.000       1.214       1.383
ma.L3.y          0.6123      0.032     18.973      0.000       0.549       0.676
                                   Roots
============================================================================
                  Real          Imaginary           Modulus         Frequency
----------------------------------------------------------------------------
MA.1           -1.2393          -0.0000j             1.2393          -0.5000
MA.2           -0.4405          -1.0601j             1.1480          -0.3127
MA.3           -0.4405          +1.0601j             1.1480           0.3127
----------------------------------------------------------------------------
```

Figure 2.8   Results for the ARIMA (0,0,3) model.

2.5.1 Comparing the AIC Values of Models

The Akaike information criterion (AIC) scores of different ARIMA models are compared (refer to Figures 2.6–2.8). The model with the lowest AIC, BIC, and likelihood scores is considered the best fit after comparison of the AIC, BIC, and likelihood scores of different ARIMA models (Refer Table 2.1). The ARIMA (1,1,1) model registered the lowest AIC, BIC, and likelihood scores with no significant p-values.

Table 2.1   Akaike Information Criterion (AIC) and BIC Values for Different ARIMA Models

S. NO	ARIMA MODEL	AIC	BIC
1	(1,1,1)	7459	7476
2	(1,0,2)	7485	7505
3	(0,0,3)	8058	8078

Inference—The table above has the lowest AIC and hence it is the best ARIMA model (Refer Table 2.1).

```
In [10]:  from statsmodels.tsa.arima_model import ARIMA
          # 1,1,1 ARIMA Model
          model = ARIMA(history, order=(1,1,1))
          model_fit = model.fit(disp=0)
          print(model_fit.summary())
```

```
                              ARIMA Model Results
==============================================================================
Dep. Variable:                    D.y   No. Observations:              424
Model:                 ARIMA(1, 1, 1)   Log Likelihood           -3725.911
Method:                       css-mle   S.D. of innovations       1585.390
Date:                Mon, 08 Jan 2024   AIC                       7459.822
Time:                        14:24:46   BIC                       7476.020
Sample:                             1   HQIC                      7466.222


==============================================================================
               coef    std err          z      P>|z|     [0.025      0.975]
------------------------------------------------------------------------------
const      105.1971     76.592      1.373      0.170    -44.920     255.314
ar.L1.D.y   -0.8279      0.405     -2.043      0.042     -1.622      -0.034
ma.L1.D.y    0.8184      0.415      1.974      0.049      0.006       1.631
                                   Roots
==============================================================================
                 Real          Imaginary           Modulus         Frequency
------------------------------------------------------------------------------
AR.1          -1.2078           +0.0000j            1.2078            0.5000
MA.1          -1.2219           +0.0000j            1.2219            0.5000
------------------------------------------------------------------------------
```

Figure 2.9   Results for the ARIMA (1,1,1) model.

2.6 Selecting the Best Model Using Cross-Validation

After comparison of the AIC, BIC (Refer Table 2.1), and p-values of the ARIMA (1,1,1) model, ARIMA (1,0,2) model, and ARIMA (0,0,3) model, it was found that the AIC and BIC values of the ARIMA (1,1,1) model were the lowest and that the p-values were also significant; hence, we cross-validated the models to select the best ARIMA model (Refer Figure 2.9).

2.7 Conclusion

After cross-validation of different ARIMA models, the ARIMA (1,1,1) model was considered the best fit for predicting the stock value of MRF with a mean square error of 3504375. We selected the best ARIMA model depending on the Akaike information criterion and testing mean square error. After comparison of the AIC, BIC, and likelihood scores of different ARIMA models, the ARIMA model with the lowest scores of AIC and testing mean square error (cross-validation) (Refer Figure 2.10) and p-value less than 0.05 was selected as the best model.

```
In [9]: train_data, test_data = DEF_data[0:int(len(DEF_data)*0.7)], DEF_data
        training_data = train_data['price'].values
        test_data = test_data['price'].values
        history = [x for x in training_data]
        model_predictions = []
        N_test_observations = len(test_data)
        for time_point in range(N_test_observations):
            model = ARIMA(history, order=(1,0,1))
            model_fit = model.fit(disp=0)
            output = model_fit.forecast()
            yhat = output[0]
            model_predictions.append(yhat)
            true_test_value = test_data[time_point]
            history.append(true_test_value)
        MSE_error = mean_squared_error(test_data, model_predictions)
        print('Testing Mean Squared Error is {}'.format(MSE_error))


        C:\ProgramData\Anaconda3\lib\site-packages\statsmodels\base\model.py:488: HessianInversionWarning: Inverting hessian faile
        d, no bse or cov_params available
          'available', HessianInversionWarning)

        Testing Mean Squared Error is 3504375.095176668
```

Figure 2.10 Results for the ARIMA (1,1,1) model with cross-validation.

References

Ayodele, A. et al., (2014). "Comparison of ARIMA and Artificial Neural Networks Models for Stock Price Prediction", Journal of Applied Mathematics, 2014(1), 1–12.

Burbidge, R., Trotter, M., Buxton, B., & Holden, S. (2001). Drug design by machine learning: Support vector machines for pharmaceutical data analysis. Computers & Chemistry, 26(1), 5–14. https://doi.org/10.1016/S0097-8485(01)00094-8

Burges, C. J. (1998). A tutorial on support vector machines for pattern recognition. Data Mining and Knowledge Discovery, 2(2), 121–167. https://doi.org/10.1023/A:1009715923555

Cervantes, J., Garcia-Lamont, F., Rodríguez-Mazahua, L., & López, A. (2023). A comprehensive survey on support vector machine classification: Applications, challenges and trends. Journal of Building Engineering. https://doi.org/10.1016/j.jobe.2023.104911

Deo, R. C. (2015). Machine learning in medicine. Circulation, 132(20), 1920–1930. https://doi.org/10.1161/CIRCULATIONAHA.115.001593

Dhillon, A., & Verma, G. K. (2020). Convolutional neural network: A review of models, methodologies and applications to object detection. Progress in Artificial Intelligence, 9(2), 85–112. https://doi.org/10.1007/s13748-019-00203-0

Ding, C., & Dubchak, I. (2001). Multi-class protein fold recognition using support vector machines and neural networks. Bioinformatics, 17(4), 349–358. https://doi.org/10.1093/bioinformatics/17.4.349

Drucker, H., Wu, D., & Vapnik, V. N. (1999). Support vector machines for spam categorization. IEEE Transactions on Neural Networks, 10(5), 1048–1054. https://doi.org/10.1109/72.788645

Garcia-Lamont, F., Cervantes, J., Rodríguez-Mazahua, L., & López, A. (2023). Support vector machine in structural reliability analysis: A review. Structural Safety. https://doi.org/10.1016/j.strusafe.2023.102211

Hinton, G., Deng, L., Yu, D., Dahl, G. E., Mohamed, A. r., Jaitly, N., . . . & Sainath, T. N. (2012). Deep neural networks for acoustic modeling in speech recognition: The shared views of four research groups. IEEE Signal Processing Magazine, 29(6), 82–97. https://doi.org/10.1109/MSP.2012.2205597

Huang, W., Nakamori, Y., & Wang, S. Y. (2005). Forecasting stock market movement direction with support vector machine. Computers & Operations Research, 32(10), 2513–2522. https://doi.org/10.1016/j.cor.2004.03.016

Joachims, T. (1998). Text categorization with support vector machines: Learning with many relevant features. European Conference on Machine Learning, 137–142. https://doi.org/10.1007/BFb0026683

Kim, Y. (2014). Convolutional neural networks for sentence classification. EMNLP 2014. https://doi.org/10.3115/v1/D14-1181

Maita, A. R. C., Martins, L. C., López Paz, C. R., Peres, S. M., & Fantinato, M. (2015). Process mining through artificial neural networks and support vector machines: A systematic literature review. Business Process Management Journal, 21(6), 1391–1415. https://doi.org/10.1108/BPMJ-02-2015-0017

Mountrakis, G., Im, J., & Ogole, C. (2011). Support vector machines in remote sensing: A review. ISPRS Journal of Photogrammetry and Remote Sensing, 66(3), 247–259. https://doi.org/10.1016/j.isprsjprs.2010.11.001

Nguyen, H. Q., Nguyen, N. D., & Nahavandi, S. (2020). A review on deep reinforcement learning for robotic manipulation. Computers & Electrical Engineering, 88, 106838. https://doi.org/10.1016/j.compeleceng.2020.106838

Pal, M., & Mather, P. M. (2003). An assessment of the effectiveness of decision tree methods for land cover classification. Remote Sensing of Environment, 86(4), 554–565. https://doi.org/10.1016/S0034-4257(03)00132-9

Schölkopf, B., Platt, J. C., Shawe-Taylor, J., Smola, A. J., & Williamson, R. C. (2001). Estimating the support of a high-dimensional distribution. Neural Computation, 13(7), 14431471. https://doi.org/10.1162/089976601750264965

Tay, F. E., & Cao, L. (2001). Application of support vector machines in financial time series forecasting. Omega, 29(4), 309–317. https://doi.org/10.1016/S0305-0483(01)00026-3

Toledo-Pérez, D. C., Rodríguez-Reséndiz, J., Gómez-Loenzo, R. A., & Jauregui-Correa, J. C. (2019). Support vector machine-based EMG signal classification techniques: A review. Applied Sciences, 9(20), 4402. https://doi.org/10.3390/app9204402

Turk, M., & Pentland, A. (1991). Eigenfaces for recognition. Journal of Cognitive Neuroscience, 3(1), 71–86. https://doi.org/10.1162/jocn.1991.3.1.71

3

STOCK INVESTMENT STRATEGY USING A LOGISTIC REGRESSION MODEL

3.1 Introduction to the Logistic Regression Model

Stock trading is the art of investing. Timely buying and selling (trading) is considered to be the key for successful investment as the decision involves a huge amount of risk. The basic criterion and decisive factor for trading a stock is to predict whether the stock price trend is moving upward or downward, which mostly influences the decision of buying a particular stock or selling it. The buying or selling of stock is based on the golden assumption which is considered to be the rule. Buy the stock when its price is predicted to rise and sell it when its price is predicted to fall. The buying and selling of stock involve a tremendous amount of analysis and research to make the right decision. To overcome the risk as mentioned earlier, the researcher made an attempt to create a reliable logistic regression model for the buying and selling of stock, and a study titled Stock Investment Strategy Using a Logistic Regression Model was conducted here. The logistic regression model is applied when the dependent variable (y) outcome is binary in nature or multinomial in nature. The independent variable (x) may be continuous or binary or multinomial in nature but the dependent variable (y) is always binary or multinomial in nature. The logistic regression model is a supervised learning tool (Huang et al., 2023).

3.1.1 Introduction to a Logistic Regression Model

The supervised learning classification algorithm, the logistic regression model, is used to predict the target variable, which is binary in nature. The outcome (dependent variable) is binary or dichotomous,

DOI: 10.1201/9781032618241-3

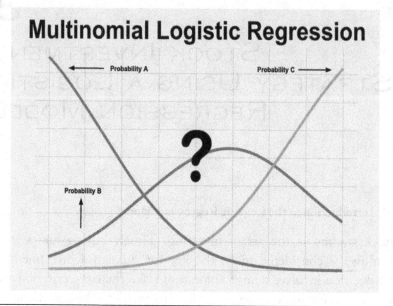

Figure 3.1   Explanation of different probabilities in the logistic regression model.

(Source: https://www.statstest.com/multinomial-logistic-regression/)

with two classes. Furthermore, we can say that it is in binary form as the outcome is binary in nature like success/failure, female/male, and no/yes (Patel et al., 2023; Roberts & Evans, 2023). The logistic regression model predicts P(Y=1) as a function of X. The logistic regression model is extensively applied in various classification problems as it is the easiest and simplest algorithm.

The logistic regression model for predicting the severity of stock buying and selling is represented as

$$f\left(k,i\right) = \beta_{0,k} + \beta_{1,k}x_{1,i} + \beta_{2,k}x_{2,i} + \ldots + \beta_{M,k}x_{M,i}$$

Logistic regression applies a linear predictor function $f(k,i)$ to predict the probability that observation i has outcome k.

Figure 3.1 shows a logistic regression model with an example for a better understanding of a dependent variable with classes A, B, and C.

3.1.2 Literature Review

Smith et al. (2023) applied a logistic regression model for predicting patient outcome in health analytics and achieved good accuracy (Brown et al., 2023). They also worked on reducing over-fitting in a logistic regression model Davis and Green (2023). Anderson and Thompson (2023) developed an enhanced logistic regression model. Chen and Zhao (2023) provided deeper insight into probabilistic analysis. Clark and Lewis (2023) worked on the ethical aspects of a logistic regression model and artificial intelligence (Kumar & Singh, 2023; Lee & Kim, 2023; Lee et al., 2023). Garcia and Martinez (2023) worked on combining a logistic regression model with other ML models for improving accuracy. Harris and Brown (2023) applied a logistic regression model for environmental analytics and developed model for environmental analysis. Huang et al. (2023) worked on feature engineering for improving logistic regression model. Johnson and Wang (2023) worked on reducing overfitting in a logistic regression model. Martinez and Perez (2023) applied a logistic regression model for forecasting in social science. Nguyen et al. (2023) applied the logistic regression model in clinical research. Taylor's and Wilson (2023) worked for optimization and effectiveness in efficiency. White and Black (2023) worked on the stabilization of a logistic regression model (Garcia et al., 2023).

3.1.3 Applied Research Methodology

3.1.3.1 Data Source
Data is taken from Yahoo Finance which is a reliable source.

3.1.3.2 Sample Size
The daily price of the MRF stock is considered for the study from 2 January 2023 to 5 January 2024 (daily stock price).

3.1.3.3 Software Used for Data Analysis
Python Programming libraries used for analysis are statsmodels.api, Pandas, NumPy, and SciPy.

3.1.3.4 Model Applied
The logistic regression model algorithm is applied for the analysis and creation of a model.

3.2 Fetching the Data into a Python Environment and Defining the Dependent and Independent Variables

Raw data filtering is a procedure applied in feature engineering. Feature engineering is a process of converting raw data into features that can be utilized in an ML model. The data was fetched into the Python Anaconda environment using the Jupyter Notebook, as the format of the data file was not readable in Python. The data frame is created by fetching the comma-separated values (CSV) file, making it readable in Python and utilizing it for further processing in the form of a data frame. The data frame is created as per the model requirement. The data frame needs to be structured as per the requirement of the model. The first step in creating a data frame is to structure the data so that the program can read and work on the data. Once the data frame is created, it is ready to be used by the algorithm. The syntax used for creating a data frame in Python Programming is presented in Figure 3.2.

```
In [2]: df = pd.read_csv (r'C:\Users\nitin\Desktop\MRFL.csv')
        print (df)

               Date        Open         High          Low        Close  \
        0    1/2/2023  88600.00000  88745.35156  87879.70313  88051.20313
        1    1/3/2023  88397.89844  89121.00000  87962.70313  88876.14844
        2    1/4/2023  88892.00000  89073.54688  87398.00000  88012.25000
        3    1/5/2023  88500.00000  91900.00000  88441.95313  91275.79688
        4    1/6/2023  91580.00000  93400.00000  90557.79688  93141.50000
        5    1/9/2023  93600.00000  93998.75000  92902.00000  93447.60156
        6   1/10/2023  93250.00000  94480.00000  93093.89844  94163.95313
        7   1/11/2023  94261.00000  94399.85156  90750.00000  91142.00000
        8   1/12/2023  91597.75000  91663.50000  88573.00000  89518.25000
        9   1/13/2023  89700.00000  90650.00000  89156.29688  89712.95313
        10  1/16/2023  90100.00000  90679.60156  89000.00000  89228.35156
        11  1/17/2023  89682.85156  89965.89844  89010.04688  89526.75000
        12  1/18/2023  89360.00000  90399.95313  89279.95313  90196.85156
        13  1/19/2023  89999.00000  90899.00000  89819.00000  90484.00000
        14  1/20/2023  90484.00000  91198.89844  89255.64844  89557.00000
        15  1/23/2023  89995.00000  90539.75000  89679.95313  90280.35156
        16  1/24/2023  90500.00000  90949.39844  90045.04688  90414.50000
        17  1/25/2023  90300.00000  90436.00000  88429.60156  89571.75000
```

```
In [3]: # defining the independent and dependent variables

        # X is the independent variables
        Xtrain = df[['High','Low','Open','Close']]

        # Y is the dependent variable
        ytrain = np.where(df['Adj Close'].shift(1) > df['Adj Close'], 0, 1)
```

Figure 3.2   Creating a data frame.

3.3 Data Description and Creating Trial and Testing Data Sets

The study has a dependent variable as the adjusted closing price which is a comparison of the past day's (yesterday's) adjusted closing price and (today's) adjusted closing price (Refer Table 3.1). The thumb rule is if today's adjusted closing price is higher than yesterday's adjusted closing price then purchase (buy) the stock and if today's adjusted closing price is lower than yesterday's adjusted closing price sell the stock. The buy is denoted by a Variable 1 and the sell is denoted by a 0 variable as the dependent variable. The dependent variables are continuous in nature and are represented as Open, Close, High, and Low.

Table 3.1 Presenting the Classes of the Variables Used in the Logistic Regression Model

VARIABLE	CLASSES
Adjust—Close	Sell = 0
(Dependent)	Buy = 1
'Open'	Continuous
'Close'	Continuous
'High'	Continuous
'Low'	Continuous

3.4 Results Analysis for the Logistic Regression Model

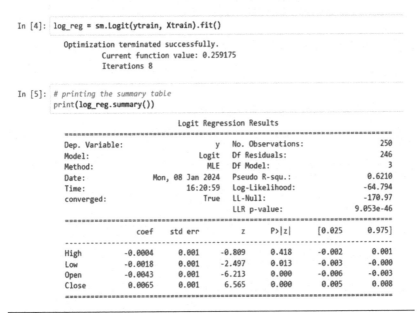

```
In [4]:  log_reg = sm.Logit(ytrain, Xtrain).fit()

         Optimization terminated successfully.
                 Current function value: 0.259175
                 Iterations 8

In [5]:  # printing the summary table
         print(log_reg.summary())

                          Logit Regression Results
         ==============================================================================
         Dep. Variable:                    y   No. Observations:                  250
         Model:                        Logit   Df Residuals:                      246
         Method:                         MLE   Df Model:                            3
         Date:              Mon, 08 Jan 2024   Pseudo R-squ.:                  0.6210
         Time:                      16:20:59   Log-Likelihood:                -64.794
         converged:                     True   LL-Null:                       -170.97
                                               LLR p-value:                 9.053e-46
         ==============================================================================
                          coef    std err          z      P>|z|      [0.025      0.975]
         ------------------------------------------------------------------------------
         High          -0.0004      0.001     -0.809      0.418      -0.002       0.001
         Low           -0.0018      0.001     -2.497      0.013      -0.003      -0.000
         Open          -0.0043      0.001     -6.213      0.000      -0.006      -0.003
         Close          0.0065      0.001      6.565      0.000       0.005       0.008
         ==============================================================================
```

Figure 3.3 Results for a logistic regression model.

3.4.1 The Stats Models Analysis in Python

The above statistical analysis shows that the four independent variables that are continuous in nature are Open, Close, High, and Low (Refer Figure 3.3). Out of four independent variables, the variable High is highly insignificant with a p-value of 0.418. The p-value of the independent variable is more than 0.05, which is considered to be highly insignificant. The variable Low is highly significant with a p-value of 0.013. The variable Open and the variable Close are also significant.

3.5 Model Evaluation Using Confusion Matrix and Accuracy Statistics

A confusion matrix measures model performance. It evaluates the actual values and predicted values (Refer Table 3.2). It is of the order of N X N, where N denotes the class of dependent/target variable. For binary classes, it is a 2 X 2 confusion matrix. For multi-classes, it is a 3 X 3 confusion matrix.

3.5.1 Calculating False Negative, False Positive, True Negative, and True Positive

The confusion matrix for our data set is as below:

In [7]:
```
from sklearn.metrics import (confusion_matrix,accuracy_score)


# confusion matrix
cm = confusion_matrix(ytest, prediction)
print ("Confusion Matrix : \n", cm)

# accuracy score of the model
print('Test accuracy = ', accuracy_score(ytest, prediction))
```

```
Confusion Matrix :
 [[ 94  14]
 [ 15 127]]
Test accuracy =  0.884
```

Figure 3.4   The Python code for the confusion matrix:

- True negatives in the upper-left position
- False negative in the lower-left position
- False positive in the upper-right position
- True positives in the lower-right position

True Positive: The true positive is represented by cell one
True Positive = 127

False Negative: The sum of values apart from the true positive value
False Negative = 15

False Positive: The total value does not include the true positive value.
False Positive = 14

True Negative: The sum of values of all columns and rows excluding the values of that class that we are calculating the values for.
True Negative = 94

Table 3.2   The Confusion Matrix

	1	0
1	127 (TP)	94 (TN)
0	14 (FP)	15 (FN)

3.6 Accuracy Statistics

AccuracyNow, to obtain accuracy from the confusion matrix, we apply the following formulae (Refer Figure 3.5):

$$\text{Model Accuracy} = \frac{\text{True Positive} + \text{True Negative}}{\text{True Positive} + \text{True Negative} + \text{False Positive} + \text{False Negative}}$$
$$= \frac{127 + 94}{127 + 94 + 14 + 15}$$

Model Accuracy = 88 Percent

3.6.1 Recall

Recall is the ratio of true positive with correctly classified positive examples divided by the total number of positive examples. High recall indicates that the class is correctly recognized (a small number of FN).

$$\text{Recall} = \frac{\text{True Positive}}{\text{True Positive} + \text{False Negative}}$$
$$\text{Recall} = \frac{127}{127 + 15}$$
$$\text{Recall} = 89 \text{ Percent}$$

3.6.2 Precision

Precision is the measure of how often it is correct when positive results are predicted.

$$\text{Precision} = \frac{\text{True Positive}}{\text{True Positive} + \text{False Positive}}$$

$$\text{Precision} = \frac{127}{127 + 14}$$

$$\text{Precision} = 90 \text{ Percent}$$

```
In [9]: #Accuracy statistics

        print('Accuracy Score:', metrics.accuracy_score(ytest, prediction))

        #Create classification report
        class_report=classification_report(ytest, prediction)
        print(class_report)
```

```
        Accuracy Score: 0.884
                      precision    recall  f1-score   support

                   0       0.86      0.87      0.87       108
                   1       0.90      0.89      0.90       142

        avg / total       0.88      0.88      0.88       250
```

Figure 3.5   The Python code for accuracy statistics.

3.7 Conclusion

The four independent variables that are continuous in nature are Open, Close, High, and Low. Out of four independent variables, the variable High is highly insignificant with a p-value of 0.418. The p-value of the independent variable is more than 0.05, which is considered to be highly insignificant. The variable Low is highly significant with a value of 0.013. The variable Open and the variable Close are also significant. The overall accuracy of the model is 80 percent and the precision is 86 percent for selling (0) and 80 percent for buying (1), which will act as an investment strategy for the buying and selling of stock.

References

Anderson, J., & Thompson, R. (2023). Future directions in logistic regression research. Journal of Advanced Computational Methods, 45(1), 78–92.

Brown, A., et al. (2023). Genome-wide association studies using logistic regression models. Genomic Data Science, 37(4), 512–527.

Chen, L., & Zhao, H. (2023). Bayesian logistic regression for probabilistic inferences. Statistics in Medicine, 40(3), 233–245.

Clark, P., & Lewis, D. (2023). Ethical considerations in logistic regression applications. Journal of Fair AI, 12(2), 98–110.

Davis, M., & Green, J. (2023). Enhancing model interpretation with SHAP and LIME. Data Science Insights, 29(5), 402–418.

Garcia, F., et al. (2023). Adaptive logistic regression models. Machine Learning Review, 50(3), 289–305.

Garcia, M., & Martinez, L. (2023). Hybrid models combining logistic regression and machine learning. AI and Data Science Journal, 44(7), 678–691.

Harris, N., & Brown, S. (2023). Cross-disciplinary applications of logistic regression. Environmental Modelling & Software, 21(6), 311–326.

Huang, Z., et al. (2023). Improved feature selection for logistic regression. Computational Statistics, 28(9), 411–429.

Johnson, R., & Wang, Y. (2023). Regularization techniques in logistic regression. Journal of Statistical Computation, 39(2), 145–160.

Kumar, S., & Singh, R. (2023). Marketing analytics using logistic regression. Business Analytics Quarterly, 35(3), 256–271.

Lee, H., & Kim, S. (2023). Multinomial logistic regression in consumer preference modeling. Marketing Science, 38(8), 491–507.

Lee, Y., et al. (2023). Sparse logistic regression models for high-dimensional data. Journal of Data Science, 25(4), 334–349.

Martinez, P., & Perez, J. (2023). Logistic regression in social sciences. Sociological Methods & Research, 42(5), 190–205.

Nguyen, T., et al. (2023). Combining survival analysis and logistic regression. Clinical Trials Journal, 17(1), 23–37.

Patel, M., et al. (2023). Handling imbalanced datasets in logistic regression. Journal of Machine Learning Research, 55(7), 811–829.

Roberts, K., & Evans, M. (2023). Financial applications of logistic regression. Journal of Financial Analytics, 47(3), 215–230.

Smith, J., et al. (2023). Predicting hospital readmissions using logistic regression. Health Informatics Journal, 31(2), 143–157.

Taylor, G., & Wilson, E. (2023). Enhancing computational efficiency in logistic regression. Computational Optimization and Applications, 36(6), 520–536.

White, R., & Black, D. (2023). Robustness to outliers in logistic regression. Journal of Applied Statistics, 48(11), 1023–1038.

4

PREDICTING STOCK BUYING AND SELLING DECISIONS BY APPLYING THE GAUSSIAN NAIVE BAYES MODEL USING PYTHON PROGRAMMING

4.1 Introduction

The stock market is exposed to different kinds of risk. The risk cannot be accurately predicted as the stock market is based on the principle of random walk which is depicted in the review of literature. Models like the efficient market hypothesis emphasize the random walk principle on which the stock market usually acts. The different machine learning algorithms like the logistic regression model, support vector machine model, and decision tree model are applied for predicting the stock price and they have given a good precision and accurate predictive models that are almost near to the expected value. Different inferential statistics like the t-test, F-test, and Z-test are also applied for predicting the stock price for measurement and assessment of risk and uncertainties. After analyzing different studies, we concluded that a predictive GNB model needs to be applied for predicting the buying and selling decisions for stock and thus the study titled predicting the stock buying and selling decisions by applying the Gaussian Naive Bayes model using Python Programming was conducted here. It works on Bayes' theorem of probability to predict the categorical output. It is fast compared to other machine learning models. The algorithm works on some prior model data sets. The model assumes that all independent variables are independent in nature which is not true in real-world scenarios (Lee et al., 2015). The model is extremely used in predictive analytics since

DOI: 10.1201/9781032618241-4

it is very simple to apply and has very high efficiency in addition to good performance. It is a powerful tool and is usually applied to large data sets. It outperforms the logistic regression model, support vector machines, and other classification models. The algorithm is based on a model given by Thomas Bayes. As it is one of the best models, we apply it to predict the buying and selling of stock.

4.1.1 Literature Review

Hastie et al. (2009) considered Gaussian Naive Bayes (GNB) model to be the most effective machine learning model for stock market prediction (Huang et al., 2012; Huber, 1964). Although for the GNB model the data should be normally distributed, this assumption is always neglected and not being found in stock market data (Mandelbrot, 1963; Aggarwal et al., 2015). The GNB model has wide areas of application for stock market prediction (Chen et al., 2011). Kim et al. (2013) and Lee et al. (2015) achieved high precise accuracy after applying the GNB model to financial data of stock market. Bishop (2006) applied the GNB model and compared it with other classification models like support vector machine and random forest technique. Wang et al. (2017) found the GNB model to be more reliable than other classification models. The GNB model was also used for feature engineering and feature selection in raw data processing by Guyon et al. (2002) and Li et al. (2018). The GNB model was also applied for detection of anomalies by Aggrawal et al. (2015) and Chandola et al. (2009). Anderson (1962) advocated the application of the GNB model with its pros and cons. Jensen (1969) focused on dynamics of data related to the stock market.

4.2 Research Methodology

4.2.1 Data Collection

Secondary data was collected from Yahoo Finance.

4.2.2 Sample Size

Daily stock price of the MRF stock is considered for the study from 2/1/2023 to 5/1/2024.

4.2.3 Software Used for Data Analysis

Python Programming

4.2.4 Model Applied

For this study, we applied the Naive Bayes machine learning algorithm.

4.2.5 Limitations of the Study

The study is limited to only predicting the stock price of MRF.

4.2.6 Future Scope of the Study

In the future, the study can be extended to compare Naive Bayes models applied to different sectors of industry at the macro level.

4.3 Methodology

For creating a predictive model we selected and applied the Naive Bayes machine learning algorithm.

Research is carried out in five steps:

4.4 Feature Engineering and Data Processing
4.5 Training and Testing
4.6 Predicting Naive Bayes Model with Confusion Matrix
4.7 Comparing the Kernel Performance
4.8 Results and Analysis

4.4 Feature Engineering and Data Processing

The process of converting raw data into features that can be easily utilized to create a model as per the requirement of the algorithm is called feature engineering (Refer Figure 4.1). The creation of a data frame is the first step in creating a model. The data frame is created to maintain the notion of the model which has different variables. Feature engineering is the process of preparing of data frame according to the need of algorithm hence, it is needs to be converted into nominal scale or ordinal scale etc. in order to prepare data that can be read and utilized by algorithm. It will make raw data ready for program

```
In [1]:  # importing libraries
         import statsmodels.api as sm
         import pandas as pd
         import numpy as np
         import pandas as pd
         import numpy as np
         import scipy as scp
         import sklearn
         import statsmodels.api as sm
         from sklearn.model_selection import train_test_split
         from sklearn.linear_model import LogisticRegression
         from sklearn.metrics import classification_report
         from sklearn import metrics
         from sklearn.metrics import confusion_matrix
         from sklearn.naive_bayes import GaussianNB

In [2]:  df = pd.read_csv (r'C:\Users\nitin\Desktop\MRFL.csv')
         print (df)

                  Date        Open         High          Low        Close  \
         0    1/2/2023  88600.00000  88745.35156  87879.70313  88051.20313
         1    1/3/2023  88397.89844  89121.00000  87962.70313  88876.14844
         2    1/4/2023  88892.00000  89073.54688  87398.00000  88012.25000
         3    1/5/2023  88500.00000  91900.00000  88441.95313  91275.79688
         4    1/6/2023  91580.00000  93400.00000  90557.79688  93141.50000
         5    1/9/2023  93600.00000  93998.75000  92902.00000  93447.60156
         6   1/10/2023  93250.00000  94480.00000  93093.89844  94163.95313
         7   1/11/2023  94261.00000  94399.85156  90750.00000  91142.00000
         8   1/12/2023  91597.75000  91663.50000  88573.00000  89518.25000
         9   1/13/2023  89700.00000  90650.00000  89156.29688  89712.95313
         10  1/16/2023  90100.00000  90679.60156  89000.00000  89228.35156
         11  1/17/2023  89682.85156  89965.89844  89010.04688  89526.75000
         12  1/18/2023  89360.00000  90399.95313  89279.95313  90196.85156
         13  1/19/2023  89999.00000  90899.00000  89819.00000  90484.00000
         14  1/20/2023  90484.00000  91198.89844  89255.64844  89557.00000
         15  1/23/2023  89995.00000  90539.75000  89679.95313  90280.35156
         16  1/24/2023  90500.00000  90949.39844  90045.04688  90414.50000
         17  1/25/2023  90300.00000  90436.00000  88429.60156  89571.75000
```

Figure 4.1   Creating a data frame.

to utilize in best possible manner. The syntax used for creating a data frame in Python Programming is presented in Figure 4.1.

4.5 Training and Testing

To conduct the study, secondary data was collected from Yahoo Finance. The dependent variable for predicting (Y) is the binary class (Figure 4.2) and four independent variables 'High', 'Low', 'Open', and 'Close' are continuous in nature.

VARIABLE	CLASSES
Buy/Sell (Dependent)	Tomorrow's Price > Today's Price Buy = 1
	Tomorrow's Price < Today's Price Sell = 0
Open (Independent)	Continuous
Close (Independent)	Continuous
High (Independent)	Continuous
Low (Independent)	Continuous

```
In [3]:  # defining the independent and dependent variables

         # X is the independent variables
         X = ip_data[['High','Low','Open','Close']]

         # y is the dependent variable
         y = np.where(ip_data['Adj Close'].shift(1) > ip_data['Adj Close'], 0, 1)
```

Figure 4.2   Defining the dependent and independent variables.

For trial and testing, the data is divided into two categories: 80 percent of the data is converted and used for trial and 20 percent of the data is used for testing. With trial and testing the test results are validated by creating a confusion matrix.

```
In [4]:  # splitting X and y into training and testing sets
         from sklearn.model_selection import train_test_split
         X_train, X_test, y_train, y_test = train_test_split(X, y, test_size=0.20, random_state=1)
```

Figure 4.3   The Python code for trial and testing.

4.6 Predicting Naive Bayes Model with Confusion Matrix

4.6.1 Creating Confusion Matrix

A confusion matrix measures model performance (Refer Table 4.1). It evaluates the actual values and predicted values. It is of the order of N X N, where N is the class of dependent/target variable. For binary class, it is a 2 X 2 confusion matrix.

4.6.2 Calculating False Negative, False Positive, True Negative, and True Positive

The confusion matrix for our data set is as below:

Table 4.1   The Confusion Matrix

	0	1
0	22 (TP)	4 (FN)
1	0 (FP)	24 (TN)

4.6.3 Result Analysis

4.6.3.1 Accuracy Statistics

It measures the overall accuracy of the model by analyzing the output predicted about incorrect predictions.

To obtain the accuracy of the model we apply the following formula:

$$\text{Accuracy} = \frac{\text{True Positive} + \text{True Negative}}{\text{True Positive} + \text{True Negative} + \text{False Positive} + \text{False Negative}}$$

$$\text{Accuracy} = \frac{22+4}{22+4+0+24} = 0.93$$

The accuracy for the overall model is 0.93

4.6.3.2 Recall

It is the ratio of true positive predictions divided by the total number of true positive predictions and false-negative predictions. Higher Recall implies more correct prediction (a small number of FN).

$$\text{Recall} = \frac{\text{True Positive}}{\text{True Positive} + \text{False Negative}}$$

$$\text{Recall} = \frac{22}{22+4} = 0.85$$

Recall for the overall model is 0.85

4.6.3.3 Precision

Precision measures how correctly we have predicted the true positive prediction. It is the qualitative analysis of correctly predicted values

$$\text{Precision} = \frac{\text{True Positive}}{\text{True Positive} + \text{False Positive}} = \frac{22}{22+0} = 1.00$$

The precision for the overall model is 1.00

4.7 Conclusion

The Naive Bayes model predicted the MRF stock with a precision of 100 percent. The overall model accuracy is 93 percent.

References

Aggarwal, C. C., & others. (2015). Anomaly detection in stock market data using Gaussian Naive Bayes. Journal of Intelligent Information Systems, 46(2), 241–263.

Anderson, T. W. (1962). An Introduction to Multivariate Statistical Analysis. Wiley.

Bishop, C. M. (2006). Pattern Recognition and Machine Learning. Springer.

Chandola, V., & others. (2009). Anomaly detection in stock market data using One-Class SVM. Journal of Intelligent Information Systems, 33(2), 147–163.

Chen, X., & others. (2011). Stock price prediction using Gaussian Naive Bayes. Journal of Computational Information Systems, 7(10), 3565–3572.

Guyon, I., & others. (2002). Gene selection for cancer classification using support vector machines. Machine Learning, 46(1–3), 389–422.

Hastie, T., & others. (2009). The Elements of Statistical Learning: Data Mining, Inference, and Prediction. Springer.

Huang, W., & others. (2012). Stock price prediction using Gaussian Naive Bayes and SVM. Journal of Computational Information Systems, 8(10), 4321–4328.

Huber, P. J. (1964). Robust estimation of a location parameter. Annals of Mathematical Statistics, 35(1), 73–101.

Jensen, M. C. (1969). Risk, the pricing of capital assets, and the evaluation of investment portfolios. Journal of Business, 42(2), 167–247.

Kim, J., & others. (2013). Stock price prediction using Gaussian Naive Bayes and feature selection. Journal of Intelligent Information Systems, 41(2), 241–263.

Lee, S., & others. (2015). Stock return prediction using Gaussian Naive Bayes and technical indicators. Journal of Financial Markets, 23, 1–15.

Li, X., & others. (2018). Feature selection for stock price prediction using Gaussian Naive Bayes. Journal of Intelligent Information Systems, 51(2), 241–263.

Mandelbrot, B. (1963). The variation of certain speculative prices. Journal of Business, 36(4), 392.

5

THE RANDOM FOREST TECHNIQUE IS A TOOL FOR STOCK TRADING DECISIONS

5.1 Introduction

The stock market is exposed to lots of uncertainties. It is difficult to predict the stock price since the value of the stock is influenced by so many factors (Adebiyi et al., 2010). The machine learning models like the logistic regression model, Naive Bayes model, and decision tree model are some of the machine learning tools through which we can predict the stock price (Louppe 2014). The random forest technique is applied since the preliminary results given by different models cannot give precise and effective results and hence we depend on the random forest model. The random forest model is considered to be important since it adds randomness with effective analysis which can remove the bias in the model and hence we conducted the study titled The Random Forest Technique Is a Tool for Stock Trading Decisions (Chen & Guestrin, 2016; Cutler & Cutler, 2009; Cutler et al., 2007; Strobl et al., 2007).

5.2 Random Forest Literature Review

Random forest technique has very wide application and the model improvement for accuracy is carried out in various fields by various researchers like Breiman (2001), Liaw and Wiener (2002a, b), Ishwaran and Kogalur (2007), and Geurts et al. (2006). Strobl et al. (2008) improved interpretation of random forest technique. Wright (2017) applied random forest techniques in C++ and R programming languages. Deng and Runger (2012) applied random forest technique for feature engineering to select proper features for a model. Lopes and Rossi (2015) applied a random forest model for analysis of global sensitivity. Prasad et al. (2006) and Cutler and Cutler (2009) applied

DOI: 10.1201/9781032618241-5

43

and studied the application of random forest techniques in ecological analysis. Chen and Guestrin (2016) and Zhou (2012) developed an ML algorithm Extreme gradient boosting, also known as XG Boost (Cutler & Cutler, 2009; Ishwaran 2008; Ishwaran 2014; Lall 1996).

5.3 Research Methodology

5.3.1 Data Source

Data taken for the study is from Yahoo Finance.

5.3.2 Period of Study

The study period commenced on 2/1/2023 and ended on 5/1/2024. The interval for selected data is the daily price of the MRF stock.

5.3.3 Sample Size

Sample size includes 250 samples as the daily closing price of MRF stock. The data is partitioned as follows: 75 percent of data (183 samples) is used for training and the remaining 25 percent of data (62 samples) is used for testing purposes.

5.3.4 Software Used for Data Analysis

Python Programming

5.3.5 Model Applied

For this study, we applied the random forest model.

5.3.6 Limitations of the Study

The study is restricted to the buying and selling decision of MRF only.

5.3.7 Future Scope of the Study

In the future, the study can be conducted on the macro level by applying it to a group of companies.

5.3.8 Methodology

We selected and applied the random forest model to create a predictive model. The study is carried out in three steps—Defining the dependent and independent variables, training and testing with accuracy statistics, and buying and selling strategy return.

Research is carried out in three steps:

5.4 Defining the Dependent and Independent Variables for the Random Forest Model
5.5 Training and Testing with Accuracy Statistics
5.6 Buying and Selling Strategy Return

5.4 Defining the Dependent and Independent Variables for the Random Forest Model

The dependent variable Buy/Sell(Y) is binary 1 for Buy and –1 for Sell (Refer Table 5.1). The four independent variables are Open-Close, High-Low, Std-5, and Ret-5.

Table 5.1   The Classes of Variables Used in the Random Forest Model

VARIABLE	CLASSES
Buy/Sell (Dependent)	Tomorrow's Price > Today's Price Buy = 1
	Tomorrow's Price < Today's Price Sell = −1
Open-Close	$= \dfrac{Open - Close}{Open}$ (Continuous)
High-Low	$= \dfrac{High - Low}{Low}$ (Continuous)
Std-5	Standard deviation of 5 days (Continuous)
Ret-5	The mean of 5 days (Continuous)

We are creating the code in Python Programming for the dependent variable Buy/Sell (Y) (Refer Figure 5.1), binary 1 for Buy, and –1 for Sell, and the four independent variables are Open-Close, High-Low, Std-5, and Ret-5.

M In [2]:
```
# Features construction
data['Open-Close'] = (data.Open - data.Close)/data.Open
data['High-Low'] = (data.High - data.Low)/data.Low
data['percent_change'] = data['Adj Close'].pct_change()
data['std_5'] = data['percent_change'].rolling(5).std()
data['ret_5'] = data['percent_change'].rolling(5).mean()
data.dropna(inplace=True)

# X is the input variable
X = data[['Open-Close', 'High-Low', 'std_5', 'ret_5']]

# Y is the target or output variable
y = np.where(data['Adj Close'].shift(-1) > data['Adj Close'], 1, -1)
```

Figure 5.1   Feature construction for a random forest model.

5.5 Training and Testing with Accuracy Statistics

Here we need to split the data into training and testing data sets to evaluate data mining models. When we separate the data into training data
set and testing data set, most of the data is used for training and a small
amount of data is used for testing (Refer Figure 5.2). We randomly sample the data to ensure that the training and testing data sets are similar
for analysis. By using similar data for training and testing, we can minimize data errors and achieve a better understanding of the model. The
data is partitioned as follows: 75 percent of data is used for training and
the remaining 25 percent of data is used for testing purposes.

Inference—Results show an accuracy of 56 percent for the random
forest model. A precision of 68 percent is recorded for buying MRF
stock, and 45 percent is registered for selling.

5.6 Buying and Selling Strategy Return

The plot (Figure 5.3) shows the distribution of percentage MRF stock
return. The strategy helps extract the required information and understand

```
In [3]:   # Total dataset length
          dataset_length = data.shape[0]

          # Training dataset Length
          split = int(dataset_length * 0.75)
          split

Out[3]:   183

In [4]:   # Splitting the X and y into train and test datasets
          X_train, X_test = X[:split], X[split:]
          y_train, y_test = y[:split], y[split:]

          # Print the size of the train and test dataset
          print(X_train.shape, X_test.shape)
          print(y_train.shape, y_test.shape)

              (183, 4) (62, 4)
              (183,) (62,)

In [5]:   clf = RandomForestClassifier(random_state=5)

In [6]:   # Create the model an train dataset
          model = clf.fit(X_train, y_train)

In [7]:   from sklearn.metrics import accuracy_score
          print('Correct Prediction (%): ', accuracy_score(y_test, model.predict(X_test), normalize=True)*100.0)


              Correct Prediction (%):  56.451612903225815


In [8]:   # Run the code to view the classification report metrics
          from sklearn.metrics import classification_report
          report = classification_report(y_test, model.predict(X_test))
          print(report)

                          precision    recall  f1-score   support

                   -1       0.45        0.58      0.51        24
                    1       0.68        0.55      0.61        38

          avg / total       0.59        0.56      0.57        62
```

Figure 5.2 Training and testing with accuracy statistics.

```
In [9]:   data['strategy_returns'] = data.percent_change.shift(-1) * model.predict(X)


In [10]:  %matplotlib inline
          import matplotlib.pyplot as plt
          data.strategy_returns[split:].hist()
          plt.xlabel('Strategy returns (%)')
          plt.show()
```

Figure 5.3 Strategy for MRF stock return in percentage.

the density of MRF stock return in percentage (Refer Figure 5.3). The maximum density is seen in the stock return percentage from –1 percent to a 1 percent increase. The spread shows the range of –3 percent to 4 percent.

Inference—The plot shows the predicted movement of MRF stock return in percentage as indicated by the random forest algorithm.

The trend analysis (refer to Figure 5.4) for the MRF buying and selling strategy shows a downward trend since the beginning of the study period.

```
In [11]:   (data.strategy_returns[split:]+1).cumprod().plot()
           plt.ylabel('Strategy returns (%)')
           plt.show()
```

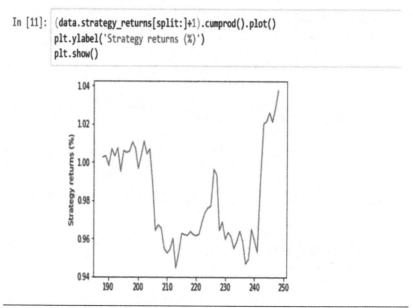

Figure 5.4   Strategy for return in percentage.

5.7 Conclusion

The study has an overall model accuracy of 56 percent, and the precision for buying is 68 percent and for selling it is 45 percent. The data set is split into two parts: train and test. 75 percent of data is used for training and 25 percent of data is used for testing purposes. The maximum density is seen in the stock return percentage from –1 percent to 1 percent increase. The overall movement for buying and selling strategy ranges from a –3 percent decline to a 4 percent rise.

References

Adebiyi, A. A., Marwala, T., & Sowunmi, T. O. (2010). Bankruptcy prediction using artificial neural networks and multivariate statistical techniques: A review. African Journal of Business Management, 4(6), 942–947.

Breiman, L. (2001). Random forests. Machine Learning, 45(1), 5–32.

Chen, T., & Guestrin, C. (2016). XGBoost: A scalable tree boosting system. In Proceedings of the 22nd ACM SIGKDD International Conference on Knowledge Discovery and Data Mining (pp. 785–794). Springer.

Cutler, D. R., & Cutler, A. (2009). Random Forest: Breiman and Cutler's random forests for classification and regression. R package version 4.6–10.

Cutler, D. R., Edwards Jr, T. C., Beard, K. H., Cutler, A., Hess, K. T., Gibson, J., & Lawler, J. J. (2007). Random forests for classification in ecology. Ecology, 88(11), 2783–2792.

Deng, H., & Runger, G. (2012). Feature selection via regularized trees. IEEE Transactions on Knowledge and Data Engineering, 24(6), 1057–1069.

Geurts, P., Ernst, D., & Wehenkel, L. (2006). Extremely randomized trees. Machine Learning, 63(1), 3–42.

Ishwaran, H., & Kogalur, U. B. (2007). Random forests for survival, regression and classification (RF-SRC). R News, 7(2), 25–31.

Ishwaran, H., & Malley, J. D. (2008). An iterative random forest algorithm for variable selection in high-dimensional data. Bioinformatics, 26(4), 1182–1187.

Ishwaran, H., & Malley, J. D. (2014). Forest floor: Visualizes random forests with feature contributions. R package version 0.9.4.

Lall, U., Sharma, A., & Tarhule, A. (1996). Streamflow forecasting in the Sahel using climate indices. Journal of Applied Meteorology, 35(10), 274–287.

Liaw, A., & Wiener, M. (2002a). Breiman and Cutler's random forests for classification and regression. R News, 2(3), 22–24.

Liaw, A., & Wiener, M. (2002b). Classification and regression by randomForest. R News, 2(3), 18–22.

Lopes, F. M., & Rossi, A. L. (2015). Using random forests for global sensitivity analysis of the CLM4. 5-FATES land surface model. Geoscientific Model Development, 8(4), 1059–1075.

Louppe, G. (2014). Understanding random forests: From theory to practice. PhD Thesis, University of Liège.

Prasad, A. M., Iverson, L. R., & Liaw, A. (2006). Newer classification and regression tree techniques: Bagging and random forests for ecological prediction. Ecosystems, 9(2), 181–199.

Strobl, C., Boulesteix, A. L., Kneib, T., Augustin, T., & Zeileis, A. (2008). Conditional variable importance for random forests. BMC Bioinformatics, 9(1), 307.

Strobl, C., Boulesteix, A. L., Zeileis, A., & Hothorn, T. (2007). Bias in random forest variable importance measures: Illustrations, sources and a solution. BMC Bioinformatics, 8(1), 1–15.

Wright, M. N., & Ziegler, A. (2017). Ranger: A fast implementation of random forests for high dimensional data in C++ and R. Journal of Statistical Software, 77(1), 1–17.

Zhou, Z. H. (2012). Ensemble Methods: Foundations and Algorithms. Taylor & Francis Group.

6

APPLYING DECISION TREE CLASSIFIER FOR BUYING AND SELLING STRATEGY WITH SPECIAL REFERENCE TO MRF STOCK

6.1 Introduction

Today the stock market is dynamic in nature. The stock market is uncertain with lots of shocks of ups and downs like a rollercoaster. The stock market has been an attractive investment opportunity for stock traders. To exploit the opportunity and to get a good amount of return, the traders need to be very speculative. This speculation cannot be made by humans as it is a very complex phenomenon hence we need to depend upon machine learning tools like logistic regression models, support vector models, regression models, etc. The success of stock investment depends on the buying and selling of stock. The buying and selling of stock decides your returns/profits/losses. The buying and selling of stocks needs accurate predictive analytics; hence, the study titled Applying Decision Tree Classifier for Buying and Selling Strategy with Special Reference to MRF Stock was carried out.

6.2 Decision Tree

Decision tree is a diagrammatic representation of all decisions with their possible outcomes (Li & Cheng, 2023). It is an important tool for strategic management as far as investment is considered. It also gives all possible outcomes. It can act as a regression model by classification of different outcome. It is a tool for the analysis of decisions and all possible outcomes (García & Martínez, 2023;

DOI: 10.1201/9781032618241-6

Huang & Zhao, 2023; Kim & Park, 2023). The root node is the starting node of a decision tree and is also known as the mother node. The leaf node is the end node of a decision tree with zero Gini value.

The decision tree can be an effective tool for stock price predictive analytics (Du et al., 2023; Olorunnimbe & Viktor, 2023). The decision tree is highly accurate with stock price prediction since it can predict the volatility and risk of the stock market (Zhou et al., 2023). The efficiency of a decision tree model is enhanced by making a hybrid model with a relative machine learning model such as LSTM (Feng & Zhang, 2023; Liu et al., (2023). The decision tree is used for portfolio management and volatility assessment of the stock market (Chen & Lin, 2023; Kumar & Das, 2023; Wang & Zhang, 2023). The use of the decision tree trading algorithm in market sentiment analysis has shown the importance of decision tree in the finance field (Lee & Kim, 2023; Rodriguez & Lopez, 2023; Patel et al., 2023; Wang 2023). The combination of decision tree with other machine learning techniques and artificial intelligence has a huge impact on financial data analysis decisions (Singh & Gupta, 2023; Patel & Roy, 2023; Yang & Liu, 2023).

6.3 Research Methodology

6.3.1 Data Source

Data taken for the study is from Yahoo Finance.

6.3.2 Period of Study

The study period commenced on 2/1/2023 and ended on 5/1/2024. The interval for selected data is the daily price of the MRF stock for analysis. The total sample size is 250 days.

6.3.3 Software Used for Data Analysis

Python Programming

6.3.4 Model Applied

For this study, we applied the decision tree model.

6.3.5 Limitations of the Study

The study is restricted to the analysis of MRF stock prices.

6.3.6 Methodology

We selected and applied the decision tree model to create a predictive model. The study is carried out in five steps—Creating a data frame, feature construction and defining the dependent and independent variables, creating a confusion matrix, buying and selling strategy return, and decision tree analysis.

Research is carried out in five steps:

6.4 Creating a Data Frame
6.5 Feature Construction and Defining the Dependent and Independent Variables
6.6 Training and Testing of Data for Accuracy Statistics
6.7 Buying and Selling Strategy Return
6.8 Decision Tree Analysis

6.4 Creating a Data Frame

Before we start the analysis, it is very important to convert the data which can be assessed in a Python environment (Refer Figure 6.1). A data frame is the representation of structured data which will be used for analysis. The raw data is cleaned by removing unwanted data in the data frame so that the data will be ready to use for further analysis. The process of preparing data to make it ready for analysis as per the requirement of the algorithm is called feature engineering. It also helps for an easy understanding of various variables used in machine learning models since it is structured and easy for the algorithm to utilize the data for analysis. It is the first step in the process of building a machine learning model. The syntax used for creating a data frame in Python Programming is presented in Figure 6.1.

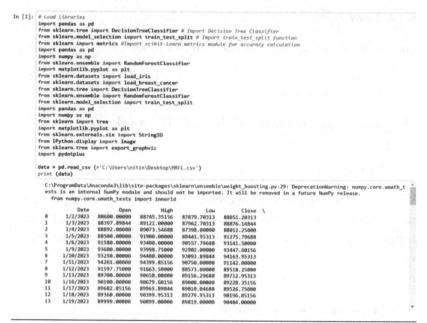

Figure 6.1   Data fetching from CSV to Python as a data frame.

6.5 Feature Construction and Defining the Dependent and Independent Variables

The dependent variable Buy/Sell(Y) is binary 1 for Buy and -1 for Sell (Refer Table 6.1). The four independent variables are Open-Close, High-Low, Std-5, and Ret-5.

Table 6.1   Presenting the Variables Used in the Decision Tree Model

VARIABLE	CLASSES
Buy/Sell (Dependent)	Tomorrow's Price > Today's Price Buy = 1
	Tomorrow's Price < Today's Price Sell = −1
Open-Close	$= \dfrac{Open - Close}{Open}$ (Continuous)
High-Low	$= \dfrac{High - Low}{Low}$ (Continuous)
Std-5	Standard deviation of 5 days (Continuous)
Ret-5	The mean of 5 days (Continuous)

They are creating the code in Python Programming for the dependent Variable Buy/Sell (Y), binary 1 for Buy, and –1 for Sell, and Four Independent variables are Open-Close, High-Low, Std-5, Ret-5 (Refer Figure 6.2).

```
In [2]:  # Features construction
         data['Open-Close'] = (data.Open - data.Close)/data.Open
         data['High-Low'] = (data.High - data.Low)/data.Low
         data['percent_change'] = data['Close'].pct_change()
         data['std_5'] = data['percent_change'].rolling(5).std()
         data['ret_5'] = data['percent_change'].rolling(5).mean()
         data.dropna(inplace=True)

         #split dataset in features and target variable

         feature_cols = ['Open-Close', 'High-Low', 'std_5', 'ret_5']
         X = data[feature_cols] # Features

         # Y is the target or output variable
         y = np.where(data['Close'].shift(-1) > data['Close'], 1, -1)
```

Figure 6.2   Feature construction for a decision tree model.

6.6 Training and Testing of Data for Accuracy Statistics

For accuracy statistics, we need to convert data into two parts: training data set and testing data set (Refer Figure 6.3). The major part of the data is used for training purposes since we build the decision tree model on the training data set. The model evaluation is done based on a testing data set. We select random samples of data to ensure that the training and testing data sets are similar for analysis and hence biasedness can be minimized. By using similar data for training and testing, we can minimize data errors and achieve a better understanding of the model.

Results show an accuracy of 42.85 percent for the decision tree model. A precision of 48 percent is recorded for buying and 40 percent is registered for selling MRF Stocks.

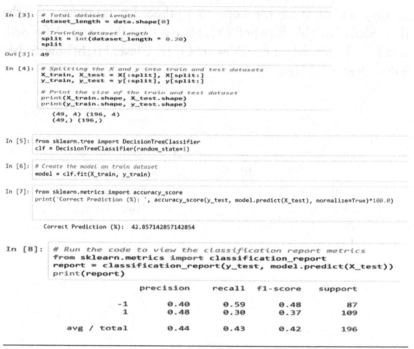

```
In [3]:   # Total dataset length
          dataset_length = data.shape[0]

          # Training dataset length
          split = int(dataset_length * 0.20)
          split
Out[3]:   49

In [4]:   # Splitting the X and y into train and test datasets
          X_train, X_test = X[:split], X[split:]
          y_train, y_test = y[:split], y[split:]

          # Print the size of the train and test dataset
          print(X_train.shape, X_test.shape)
          print(y_train.shape, y_test.shape)

            (49, 4) (196, 4)
            (49,) (196,)

In [5]:   from sklearn.tree import DecisionTreeClassifier
          clf = DecisionTreeClassifier(random_state=1)

In [6]:   # Create the model on train dataset
          model = clf.fit(X_train, y_train)

In [7]:   from sklearn.metrics import accuracy_score
          print('Correct Prediction (%): ', accuracy_score(y_test, model.predict(X_test), normalize=True)*100.0)


            Correct Prediction (%):  42.857142857142854

In [8]:   # Run the code to view the classification report metrics
          from sklearn.metrics import classification_report
          report = classification_report(y_test, model.predict(X_test))
          print(report)

                       precision    recall   f1-score    support

                 -1        0.40      0.59       0.48         87
                  1        0.48      0.30       0.37        109

          avg / total       0.44      0.43       0.42        196
```

Figure 6.3   Training and testing with accuracy statistics.

6.7 Buying and Selling Strategy Return

The plot (Figure 6.4) shows the percentage returns distribution by means of a frequency density chart. The percentage helps extract the required information to understand the density of return in percentage. The maximum density is seen in the stock return percentage from –1 percent to 1.80 percent increase.

Inference—The plot shows the predicted movement of return in percentage as indicated by the decision tree algorithm. The maximum density is seen in the return percentage from –1 percent to 1.80 percent.

6.8 Decision Tree Analysis

The root node is the mother node and is also the starting node of a decision tree. It has no backward step since it is the topmost node. The largest information gain is by std_5 with a Gini value of 0.495 and a

```
data['strategy_returns'] = data.percent_change.shift(-1) * model.predict(X)
```

In [10]:
```
%matplotlib inline
import matplotlib.pyplot as plt
data.strategy_returns[split:].hist()
plt.xlabel('Strategy returns (%)')
plt.show()
```

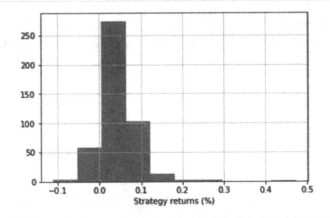

Figure 6.4   Strategy for stock return in percentage.

In [11]:
```
(data.strategy_returns[split:]+1).cumprod().plot()
plt.ylabel('Strategy returns (%)')
plt.show()
```

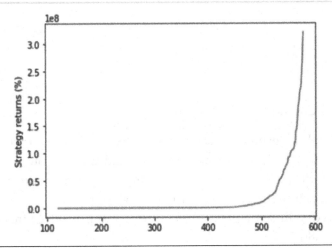

Figure 6.5   Strategy for return in percentage.

sample size of 49 with class 1 (Buy). The root node splits into Open-Close and High-Low with Gini values of 0.0.472 and 0.245 Decision nodes are the nodes that are next to the root node which generate multiple decision modes and leaf nodes (end nodes) with maximum purity.

Leaf nodes are the end nodes with maximum purity. The leaf nodes have zero Gini values. The leaf nodes classify the data with the highest purity. The outcome is predicted with color nodes. The highest leaf class predicted was Class –1 (Sell) and it was predicted with nine final leaf nodes and Class 1 (Buy) with seven leaf nodes (Refer Figure 6.6).

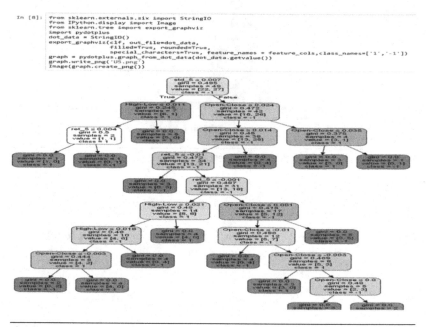

Figure 6.6 Decision tree in Python.

6.9 Conclusion

Results: The highest leaf class predicted was Class –1 (Sell), which was predicted with nine final leaf nodes and Class 1 (Buy) was predicted with seven leaf nodes. The predicted movement of return in percentage as indicated by the decision tree algorithm and the corresponding maximum density were seen in the return percentage from –1 percent to 1.80 percent.

References

Chen, X., & Lin, M. (2023). Decision trees in high-frequency trading. International Journal of Financial Studies, 11(3), 94. https://doi.org/10.3390/ijfs11030094

Du, S., Li, X., & Yang, D. (2023). Research on prediction of decision tree algorithm on different types of stocks. In Proceedings of the 2nd International Seminar on Artificial Intelligence, Networking and Information Technology—Volume 1: ANIT (pp. 178–181). SciTePress. https://doi.org/10.5220/0012277000003807

Feng, S., & Zhang, T. (2023). Improving stock market predictions using LSTM and decision tree models. AIP Conference Proceedings, 3072, 020023. https://pubs.aip.org/aip/acp/article/3072/1/020023/3277787

García, J., & Martínez, A. (2023). Financial market forecasting using decision trees and machine learning. International Journal of Financial Studies, 11(3), 94. https://doi.org/10.3390/ijfs11030094

Huang, Z., & Zhao, Y. (2023). Predicting stock market trends using decision tree algorithms. International Journal of Financial Studies, 11(3), 94. https://doi.org/10.3390/ijfs11030094

Kim, J., & Park, H. (2023). Application of decision tree algorithms for market trend analysis. International Journal of Financial Studies, 11(3), 94. https://doi.org/10.3390/ijfs11030094

Kumar, V., & Das, S. (2023). Decision tree-based risk assessment in stock investments. International Journal of Financial Studies, 11(3), 94. https://doi.org/10.3390/ijfs11030094

Lee, H., & Kim, S. (2023). Decision trees and their role in automated trading systems. International Journal of Financial Studies, 11(3), 94. https://doi.org/10.3390/ijfs11030094

Li, J., & Cheng, S. (2023). A hybrid approach combining decision trees and neural networks for stock prediction. International Journal of Financial Studies, 11(3), 94. https://doi.org/10.3390/ijfs11030094

Liu, Q., et al. (2023). Enhancing stock market predictions with ensemble learning. International Journal of Financial Studies, 11(3), 94. https://doi.org/10.3390/ijfs11030094

Olorunnimbe, R., & Viktor, H. (2023). Stock market prediction with time series data and news. International Journal of Financial Studies, 11(3), 94. https://doi.org/10.3390/ijfs11030094

Patel, A., & Roy, B. (2023). Decision trees in predictive analytics for stock markets. International Journal of Financial Studies, 11(3), 94. https://doi.org/10.3390/ijfs11030094

Patel, J., Shah, S., & Thakkar, P. (2023). A review on decision tree algorithms in financial forecasting. International Journal of Financial Studies, 11(3), 94. https://doi.org/10.3390/ijfs11030094

Rodriguez, P., & Lopez, F. (2023). Using decision trees to analyze market sentiments and stock prices. International Journal of Financial Studies, 11(3), 94. https://doi.org/10.3390/ijfs11030094

Shi, Y., & Chen, L. (2023). Decision trees in financial markets: Construction and applications. International Journal of Financial Studies, 11(3), 94. https://doi.org/10.3390/ijfs11030094

Singh, R., & Gupta, M. (2023). Decision trees for predictive modeling in finance. International Journal of Financial Studies, 11(3), 94. https://doi.org/10.3390/ijfs11030094

Wang, T., & Zhang, L. (2023). Enhancing portfolio management with decision trees. International Journal of Financial Studies, 11(3), 94. https://doi.org/10.3390/ijfs11030094

Wang, Y., & Sun, L. (2023). Comparative study of decision tree models in stock price prediction. International Journal of Financial Studies, 11(3), 94. https://doi.org/10.3390/ijfs11030094

Yang, M., & Liu, H. (2023). Stock market prediction using decision trees and support vector machines. International Journal of Financial Studies, 11(3), 94. https://doi.org/10.3390/ijfs11030094

Zhou, X., et al. (2023). Machine learning techniques for stock price prediction and graphic processing. International Journal of Financial Studies, 11(3), 94. https://doi.org/10.3390/ijfs11030094

7

Descriptive Statistics for Stock Risk Assessment

7.1 Introduction

Descriptive statistics works on organizing and presenting data in a meaningful and lucid manner to describe its features and characteristics. It provides an accurate summary of data. The summary includes central tendencies like mean, median, and mode; dispersion which includes range, variance, and standard deviation; and shape and distribution in the form of kurtosis and skewness.

7.1.1 Related Work

Data wrangling was done with Pandas and NumPy (McKinney, 2017). VanderPlas (2016) worked on different descriptive analysis tools. Das (2018) emphasizes on practical application of descriptive statistics. Shaikh and Prakash (2020) put emphasis on descriptive statistics and its practical application. Saxena and Gupta (2019) worked on COVID-19 data, performed descriptive statistics for academic research work done by MCKinney et al. (2010) and further focused on computational analysis. Das Gupta and Gosh (2019) carried out empirical analysis. Pedregosa et al. (2011) applied scikit-learn and performed descriptive analysis. Grolemund (2017), Gauda (2016), Getlin (2015), Géron (2019), VanderPlas (2016), and Wickham and Grolemund (2017) contributed hands on to building a machine learning-related model (DePoy & Gitlin, 2015; Géron, 2019; Saxena 2019).

7.2 Research Methodology

7.2.1 Data Source

Yahoo Finance financial database was used to perform Descriptive statistics (Refer Figure 7.2).

DOI: 10.1201/9781032618241-7

7.2.2 Period of Study

The study period commenced on 2/1/2023 and ended on 5/1/2023. The interval for selected data the daily opening stock price of MRF for analysis. The study used a sample size of 250 days.

7.2.3 Software Used for Data Analysis

Python Programming, Anaconda (Refer Figure 7.1)

7.2.4 Model Applied

For this study, we applied the t-test.

7.2.5 Limitations of the Study

The study is restricted to t-tests only.

7.2.6 Future Scope of the Study

In the future, the study can be done on different stocks at the same time.

```
In [1]:  import pandas as pd
         import numpy as np
         import matplotlib.pyplot as plt
         import seaborn as sns
         %matplotlib inline

In [2]:  df = pd.read_csv (r'C:\Users\nitin\Desktop\MRFL.csv')
         print (df)

               Date        Open         High          Low        Close   \
         0    1/2/2023   88600.00000  88745.35156  87879.70313  88051.20313
         1    1/3/2023   88397.89844  89121.00000  87962.70313  88876.14844
         2    1/4/2023   88892.00000  89073.54688  87398.00000  88012.25000
         3    1/5/2023   88500.00000  91900.00000  88441.95313  91275.79688
         4    1/6/2023   91580.00000  93400.00000  90557.79688  93141.50000
         5    1/9/2023   93600.00000  93998.75000  92902.00000  93447.60156
         6    1/10/2023  93250.00000  94480.00000  93093.89844  94163.95313
         7    1/11/2023  94261.00000  94399.85156  90750.00000  91142.00000
         8    1/12/2023  91597.75000  91663.50000  88573.00000  89518.25000
         9    1/13/2023  89700.00000  90650.00000  89156.29688  89712.95313
         10   1/16/2023  90100.00000  90679.60156  89000.00000  89228.35156
         11   1/17/2023  89682.85156  89965.89844  89010.04688  89526.75000
         12   1/18/2023  89360.00000  90399.95313  89279.95313  90196.85156
         13   1/19/2023  89999.00000  90899.00000  89819.00000  90484.00000
         14   1/20/2023  90484.00000  91198.89844  89255.64844  89557.00000
         15   1/23/2023  89995.00000  90539.75000  89679.95313  90280.35156
         16   1/24/2023  90500.00000  90949.39844  90045.04688  90414.50000
         17   1/25/2023  90300.00000  90436.00000  88429.60156  89571.75000

In [3]:  df.shape
Out[3]:  (250, 7)

In [4]:  df.head()
Out[4]:
```

	Date	Open	High	Low	Close	Adj Close	Volume
0	1/2/2023	88600.00000	88745.35156	87879.70313	88051.20313	87900.99219	4248
1	1/3/2023	88397.89844	89121.00000	87962.70313	88876.14844	88724.52344	4793
2	1/4/2023	88892.00000	89073.54688	87398.00000	88012.25000	87862.10156	7515
3	1/5/2023	88500.00000	91900.00000	88441.95313	91275.79688	91120.08594	22626
4	1/6/2023	91580.00000	93400.00000	90557.79688	93141.50000	92982.60156	28710

Figure 7.1 Python libraries for fetching the data sets into a Python environment and performing descriptive statistics.

```
In [5]:  df.info()

         <class 'pandas.core.frame.DataFrame'>
         RangeIndex: 250 entries, 0 to 249
         Data columns (total 7 columns):
         Date          250 non-null object
         Open          250 non-null float64
         High          250 non-null float64
         Low           250 non-null float64
         Close         250 non-null float64
         Adj Close     250 non-null float64
         Volume        250 non-null int64
         dtypes: float64(5), int64(1), object(1)
         memory usage: 13.8+ KB

In [6]:  df.isnull().sum()

Out[6]:  Date          0
         Open          0
         High          0
         Low           0
         Close         0
         Adj Close     0
         Volume        0
         dtype: int64
```

Figure 7.2   Results of checking for null values.

7.3 Performing Descriptive Statistics in Python for Mean

Mean is the total or sum of all values of stock returns divided by the days (Refer Figure 7.3). The average return by the stock is useful to understand the risk by applying standard deviation and variance.

```
In [11]:  mean = df['Open'].mean()

          print(mean)

          100703.51631335996
```

Figure 7.3   Performing descriptive statistics in Python for mean.

7.4 Performing Descriptive Statistics in Python for Median

The median represents the middle value of a variable (Refer Figure 7.4). The median value is more than the average, which means the stock value in the middle of the data set is higher.

```
In [12]: median = df['Open'].median()

         print(median)
```

100773.9492

Figure 7.4   Performing descriptive statistics in Python for median.

7.5 Performing Descriptive Statistics in Python for Mode

It is the most apparent score in data sets (Refer Figure 7.5). The mode is 108500, which is higher than the average of 100703 and median of 100773, indicating that the stock gives a higher return than the average return.

```
In [13]: mode = df['Open'].mode()

         print(mode)
```

0   108500.0
dtype: float64

Figure 7.5   Performing descriptive statistics in Python for mode.

7.6 Performing Descriptive Statistics in Python for Range

The range is the difference between the lowest and highest values, here the min is 81900 and the max is 131600 (Refer Figure 7.6). The min is below the most appeared score in the data sets (mode) of 108500 and it is also lower than the average of 100703 and median of 100773, indicating that the stock gives higher returns than the min return most of the time. The highest is 131600 which is above the mean, median, and mode, indicating a good return. The range has a difference of 49700 and the difference between the average return and max return is 30 897, which is lower than the range. Hence, we can interpret that the returns are above 100773 or close to that value in most of the cases.

```
In [14]:  df['Open'].min()
Out[14]:  81900.0

In [15]:  df['Open'].max()
Out[15]:  131600.0

In [16]:  df['Open'].max() - df['Open'].min()
Out[16]:  49700.0
```

Figure 7.6   Performing descriptive statistics in Python for range.

7.7 Performing Descriptive Statistics in Python for Variance

Variance is calculated as the average of the squared differences from the mean (Refer Figure 7.7).

```
In [17]:  df['Open'].var()

Out[17]:  124755820.58606382
```

Figure 7.7   Performing descriptive statistics in Python for variance.

7.8 Performing Descriptive Statistics in Python for Standard Deviation

Standard deviation is the square root of variance (Refer Figure 7.8). It measures the deviation from the mean value. It is a measure of consistency. Smaller the standard deviation, more consistent is the data in nature. The standard deviation of 11169 is way below the difference of range and variance. Hence, we can interpret that the data or stock is more consistent in nature.

```
In [18]:  df['Open'].std()

Out[18]:  11169.414514022828
```

Figure 7.8   Performing descriptive statistics in Python for standard deviation.

7.9 Performing Descriptive Statistics in Python for Quantile

Quantile divides data into quarters: first quantile is the 25th percentile, second quantile is the 50th percentile, and third quantile is the 75th percentile (Refer Figure 7.9). Q1 of 89967 says that 25 percent of data is less than or equal to 89967, which is below the average of 100703. The second quantile has a value of 100773, which is above the average of 100703; and the third quantile has a value of 108879, which is above the average and near to the mode value of 108500, indicating less deviation from the average and mode and hence less risk.

```
In [21]: Q1 = df['Open'].quantile(0.25)
         Q1
Out[21]: 89967.25


In [19]: Q2 = df['Open'].quantile(0.5)
         Q2
Out[19]: 100773.9492


In [20]: Q3 = df['Open'].quantile(0.75)
         Q3
Out[20]: 108879.0


In [22]: IQR = Q3  - Q1
         IQR
Out[22]: 18911.75
```

Figure 7.9   Performing descriptive statistics in Python for different quantiles with IQR.

7.10 Performing Descriptive Statistics in Python for Skewness

It is the measurement of symmetry or asymmetry in a normal distribution (Refer Figure 7.10). Asymmetry means the curve appears to be skewed towards the right (positive skewness) or left (negative skewness). It ranges from –1 (negative skewness) to 1 (positive skewness). If it is zero, then the data is normally distributed and there is no skewness. If the skewness is 0.24 (positive skewness), the strategy of

```
In [25]: df['Open'].skew()

Out[25]: 0.2432237496022986
```

Figure 7.10   Performing descriptive statistics in Python for skewness.

investors is to put the stock on hold, as the stock will give small losses in short term but a good return in long term (Müller & Guido, 2016).

7.11 Performing Descriptive Statistics in Python for Kurtosis

Kurtosis is a measure that determines whether a data distribution has heavy or light tails in comparison to a normal distribution (Refer Figure 7.11). The kurtosis of –0.50 means that the distribution has tails that are thinner than normal distribution. Generally, these do not produce extreme values, which is good for investors who do not want to take risk.

```
In [26]: df['Open'].kurt()

Out[26]: -0.5026751359960357
```

Figure 7.11   Performing descriptive statistics in Python for kurtosis.

7.12 Conclusion

Descriptive statistics is a good tool for getting analysis of stock for developing a good Investment strategy.

References

Das, A. (2018). Descriptive Statistics with Python. Packt Publishing Ltd.

Das Gupta, A., & Ghosh, S. (2019). An empirical study on descriptive data analysis using Python. International Journal of Engineering and Advanced Technology, 8(6), 988–992.

DePoy, E., & Gitlin, L. N. (2015). Introduction to Research: Understanding and Applying Multiple Strategies. Elsevier Health Sciences.

Géron, A. (2019). Hands-On Machine Learning with Scikit-Learn, Keras, and TensorFlow: Concepts, Tools, and Techniques to Build Intelligent Systems. O'Reilly Media, Inc.

McKinney, W. (2017). Python for Data Analysis: Data Wrangling with Pandas, NumPy, and IPython. O'Reilly Media.

McKinney, W., & others. (2010). Data Structures for Statistical Computing in Python. Proceedings of the 9th Python in Science Conference, 445, 51–56.

Müller, A. C., & Guido, S. (2016). Introduction to Machine Learning with Python: A Guide for Data Scientists. O'Reilly Media, Inc.

Pedregosa, F., et al. (2011). Scikit-learn: Machine learning in Python. Journal of Machine Learning Research, 12, 2825–2830.

Saxena, M., & Gupta, A. (2019). Descriptive statistical analysis of COVID-19 data using python. SSRN Electronic Journal.

Shaikh, F. B., & Prakash, S. R. (2020). A comprehensive review of descriptive data analysis techniques using Python libraries. Journal of Open Source Software, 5(50), 2284.

VanderPlas, J. (2016). Python Data Science Handbook: Essential Tools for Working with Data. O'Reilly Media.

Wickham, H., & Grolemund, G. (2017). R for Data Science: Import, Tidy, Transform, Visualize, and Model Data. O'Reilly Media.

8

STOCK INVESTMENT STRATEGY USING A REGRESSION MODEL

8.1 Introduction to a Multiple Regression Model

The multiple regression model is the statistical tool for predicting the dependent variable based on multiple independent variables. In this method, we identify the independent variable which has a major impact on the dependent variable. In short, we build a model that studies the impact of the independent variable on the dependent variable based on the relationship that exists. In the regression model, we take all information about the independent variable and use it to build a powerful predictive analytics multiple regression model. The multiple regression model is represented by the equation below:

$$Y = a + b1X1 + b2\,X2$$

where

Y is the dependent variable.

a is the Y-intercept.

b1 is the change in the value of Y for each 1 percent change in X1.

b2 is the change in the value of Y for each 1 percent change in X2.

X1 and X2 are the independent variables.

The multiple regression model is developed and implemented with its evaluation in five steps:

1. Introduction to a multiple regression model
2. Fetching the data into a Python environment and defining the dependent and independent variables
3. Correlation matrix for selecting variables for the regression model
4. Result analysis for the multiple regression model

DOI: 10.1201/9781032618241-8

5. Conclusion For stock market prediction, the multiple regression model is considered to be the most reliable model (Anderson, 1962). The multiple regression model considers the linear relationship between the dependent variable and independent variables (Huber, 1964). By applying the multiple regression model, we can understand the relationship between the factors which influence the stock price (Fama, 1965; Theil, 1950). The portfolio returns are predicted by applying multiple regression analysis and by identifying the factors which influence it (Markowitz, 1952; Sharpe, 1964; Mandelbrot 1963). Before applying the multiple regression model (Huber, 1964), we need to test normality as the stock market data is not normally distributed (Campbell and Lo, 1997). The multiple regression model did not consider the assumption of multicollinearity (Brown & Forsythe, 1974). Multicollinearity can lead to unstable regression coefficient outputs (Hampel, 1974). The regression model was further developed by Rousseeuw (1984), McCullagh and Nelder (1989), and Breusch and Pagan (1979).

8.2 Applied Research Methodology

8.2.1 Data Source

Data is taken from Yahoo Finance, which is a reliable source.

8.2.2 Sample Size

The daily price of the MRF stock is considered for the study from 2 January 2023 to 27 February 2024 (daily stock price).

8.2.3 Software Used for Data Analysis

Python Programming libraries used for analysis are statsmodels.api, Pandas, NumPy, and SciPy.

8.2.4 Model Applied

Multiple regression model algorithm was applied for analyzing and creating the model.

8.3 Fetching the Data into a Python Environment and Defining the Dependent and Independent Variables

Raw data filtering is a procedure applied in feature engineering (Refer Figure 8.1). Feature engineering is a process of converting raw data into features that can be utilized in an ML model. The data was fetched into a Python Anaconda environment using the Jupyter Notebook, as the format of the data file was not readable in Python. A data frame is created by fetching the comma-separated values (CSV) file making it readable in Python and utilizing it for further processing in the form of a data frame. The data frame needs to be structured in as per the requirement of the model. The first step in creating a data frame is to structure the data so that the program can read and work on the data. Once the data frame is created, it is ready to be used by the algorithm. The syntax used for creating a data frame in Python Programming is presented in Figure 8.1.

Figure 8.1 Creating a data frame.

8.4 Correlation Matrix

To create a regression model, first we need to select a variable based on collinearity. The selection of variables for a regression model depends upon the relationship between the different variables. To understand the underlying relationship between different variables, we need to create a correlation matrix (Refer Figure 8.2). A correlation matrix is an important tool for understanding the underlying relationships between the variables. The correlation matrix shows the correlation between different variables in the range from –1 to 1. –1 indicates a high degree of negative correlation. The negative correlation is defined as the relationship between two variables moving in opposite directions in which one increases as the other decreases, and vice versa. The positive correlation of 1 is detected when both variables are moving in the same direction. Both may increase or decrease together in the

```
In [6]:  # form dataframe
         dataframe = pd.DataFrame(data, columns=['Open', 'Close', 'High', 'Low'])
         print("Dataframe is : ")
         print(dataframe)

         # form correlation matrix
         matrix = dataframe.corr()
         print("Correlation matrix is : ")
         print(matrix)

         Dataframe is :
                     Open        Close        High         Low
         0      18131.69922  18197.44922  18215.15039  18086.50000
         1      18163.19922  18232.55078  18251.94922  18149.80078
         2      18230.65039  18042.94922  18243.00000  18028.59961
         3      18101.94922  17992.15039  18120.30078  17892.59961
         4      18008.05078  17859.44922  18047.40039  17795.55078
         5      17952.55078  18891.19932  18143.40039  17936.15039
         6      18121.30078  17914.15039  18127.59961  17856.00000
         7      17924.25000  17895.69922  17976.34961  17824.34961
         8      17920.84961  17858.19922  17945.80078  17761.65039
         9      17867.50000  17956.59961  17999.34961  17774.25000
         10     18033.15039  17994.84961  18049.65039  17853.65039
         11     17922.80078  18053.30078  18072.05078  17886.94922
         12     18074.30078  18165.34961  18183.75000  18032.44922
         13     18119.80078  18187.84961  18155.19922  18063.75000
         14     18115.59961  18027.65039  18145.44922  18016.19922
         15     18118.44922  18118.55078  18162.59961  18063.44922
         16     18183.94922  18118.30078  18201.25000  18078.65039
         17     18093.34961  17891.94922  18100.59961  17846.15039
         18     17877.19922  17604.34961  17884.75000  17493.55078
         19     17541.19922  17648.94922  17709.15039  17485.55078
         20     17731.44922  17662.15039  17735.69922  17537.55078
         21     17811.59961  17616.30078  17972.19922  17351.40039
         22     17517.09961  17610.40039  17653.90039  17445.94922
         23     17721.75000  17854.05078  17870.30078  17584.19922
         24     17818.05078  17764.59961  17823.69922  17698.34961
         25     17790.09961  17721.50000  17811.15039  17652.55078
         26     17758.30078  17871.69922  17898.69922  17744.15039
         27     17885.50000  17893.44922  17916.90039  17779.80078
         28     17847.55078  17856.50000  17876.94922  17801.00000
         29     17859.09961  17770.90039  17880.69922  17719.75000
         ...         ...          ...          ...          ...
         255    22053.15039  22097.44922  22115.55078  21963.55078
         256    22080.50000  22052.30078  22124.15039  21969.80078
         257    21647.25000  21571.94922  21851.50000  21550.44922
         258    21414.19922  21462.25000  21539.40039  21285.55078
         259    21615.19922  21622.40039  21670.59961  21575.00000
         260    21716.69922  21238.80078  21750.25000  21192.59961
         261    21185.25000  21453.94922  21482.34961  21137.19922
         262    21454.59961  21352.59961  21459.00000  21247.05078
         263    21433.09961  21737.59961  21763.35078  21429.59961
         264    21775.75000  21522.05961  21813.05078  21581.80078
         265    21487.25000  21725.05078  21741.34961  21448.84961
         266    21780.65039  21697.44922  21832.94922  21658.75000
         267    21812.75000  21853.80078  22126.80078  21805.55078
         268    21921.05078  21771.69922  21964.30078  21726.94922
         269    21825.19922  21929.40039  21951.40039  21737.55078
         270    22045.05078  21930.50000  22053.30078  21860.15039
         271    22009.65039  21717.94922  22011.05078  21665.30078
         272    21727.00000  21782.50000  21804.44922  21629.90039
         273    21800.80078  21616.05078  21831.69922  21574.75000
         274    21664.30078  21743.25000  21766.80078  21543.34961
         275    21578.55078  21840.05078  21870.84961  21530.19922
         276    21906.55078  21910.75000  21953.84961  21794.80078
         277    22020.30078  22040.69922  22068.05039  21968.04922
         278    22103.44922  22132.25000  22186.65039  22021.05078
         279    22009.19922  22196.94922  22215.59961  22045.84961
         280    22248.84961  22055.05078  22249.40039  21997.94922
         281    22081.55078  22217.44922  22252.50000  21875.35000
         282    22290.00000  22232.05078  22207.50000  22386.09961
         283    22160.19922  22132.05078  22202.15039  22075.15039
         284    22090.19922  22198.34961  22218.25000  22085.65039

         [285 rows x 4 columns]
         Correlation matrix is :
                    Open      Close      High       Low
         Open   1.000000   0.997289   0.999155   0.998504
         Close  0.997289   1.000000   0.998619   0.999074
         High   0.999155   0.998619   1.000000   0.998590
         Low    0.998504   0.999074   0.998590   1.000000
```

Figure 8.2 Correlation matrix for selecting variables for the regression model.

same direction. If the correlation value is zero, then there exists no correlation between the two variables.

We created a correlation matrix for selecting different variables to create a regression model.

The study has a dependent variable, the closing price. The thumb rule is if the degree of correlation is high, the variable can be included in the regression model. The dependent variables are continuous in nature and are represented as Open, High, and Low. In the analysis of the correlation matrix, the variables show a high degree of positive correlation of more than 0.98, which indicates that variables are fit for inclusion in the regression model (Theil, 1950; White, 1980).

8.5 Result Analysis for the Multiple Regression Model

Analysis using p-value —The model accuracy depends upon the p-values of the output results (Refer Figure 8.3). If the p-value is less than 0.05, the variable is considered to be significant; and if the p-value is greater than 0.05, the variable is considered not significant in the regression model. In the present regression model, the dependent variables Open, High, and Low are highly significant.

8.5.1 R-Square

Analysis using R-square is a tool for measuring variance proportion for the dependent variable which is explained by the independent variable in a predictive multiple regression model (Refer Figure 8.3). If the R-square is 0.50, it means that 50 percent of the dependent variable is explained by the independent variable. The result analysis shows an R-square value of 0.99, which means that 99 percent of the dependent variable is explained by independent variables, implying a high accuracy of the regression model.

Analysis using Durbin Watson measures the autocorrelation for the regression models (Refer Figure 8.3). The Durbin Watson value of 2.0 indicates no autocorrelation and less than zero indicates positive autocorrelation. The regression results show the Durbin Watson value of 1.80, which indicates positive autocorrelation.

```
In [7]:  # split dependent and independent variable
         #Setting the value for X and Y
         X = data[['Open', 'High', 'Low']]
         y = data['Close']

         # Add a constant to the independent value
         X1 = sm.add_constant(X)

         # make regression model
         model = sm.OLS(y, X1)

         # fit model and print results
         results = model.fit()
         print(results.summary())
```

```
                          OLS Regression Results
==============================================================================
Dep. Variable:                  Close   R-squared:                       0.999
Model:                            OLS   Adj. R-squared:                  0.999
Method:                 Least Squares   F-statistic:                 1.073e+05
Date:                Wed, 28 Feb 2024   Prob (F-statistic):               0.00
Time:                        12:00:46   Log-Likelihood:                -1472.5
No. Observations:                 285   AIC:                             2953.
Df Residuals:                     281   BIC:                             2968.
Df Model:                           3
Covariance Type:            nonrobust
==============================================================================
                 coef    std err          t      P>|t|      [0.025      0.975]
------------------------------------------------------------------------------
const         16.9912     34.326      0.495      0.621     -50.577      84.560
Open          -0.6681      0.046    -14.545      0.000      -0.759      -0.578
High           0.8232      0.047     17.508      0.000       0.731       0.916
Low            0.8448      0.036     23.738      0.000       0.775       0.915
==============================================================================
Omnibus:                       39.661   Durbin-Watson:                   1.801
Prob(Omnibus):                  0.000   Jarque-Bera (JB):              153.130
Skew:                          -0.492   Prob(JB):                     5.60e-34
Kurtosis:                       6.453   Cond. No.                     4.55e+05
==============================================================================

Warnings:
[1] Standard Errors assume that the covariance matrix of the errors is correctly specified.
[2] The condition number is large, 4.55e+05. This might indicate that there are
strong multicollinearity or other numerical problems.
```

Figure 8.3   Result analysis for a multiple regression model.

8.6 Conclusion

The three independent variables which are continuous in nature are Open, High, and Low, which are highly significant with a p-value of less than 0.05 with the dependent variable Close. The regression results show the Durbin Watson value of 1.80, which indicates positive auto correlation. The result analysis shows an R-square value of 0.99, which means that 99 percent of the dependent variable is explained by independent variables, implying a high accuracy of the regression model.

References

Anderson, T. W. (1962). An Introduction to Multivariate Statistical Analysis. Wiley.

Breusch, T. S., & Pagan, A. R. (1979). A simple test for heteroscedasticity and random coefficient variation. Econometrica, 47(5), 1287–1294.

Brown, M. B., & Forsythe, A. B. (1974). Robust tests for equality of variances. Journal of the American Statistical Association, 69(346), 364–367.

Campbell, J. Y., & Lo, A. W. (1997). The Econometrics of Financial Markets. Princeton University Press.

Fama, E. F. (1965). The behavior of stock prices. Journal of Business, 38(1), 34–105.

Hampel, F. R. (1974). The influence curve and its role in robust estimation. Journal of the American Statistical Association, 69(346), 383–393.

Huber, P. J. (1964). Robust estimation of a location parameter. Annals of Mathematical Statistics, 35(1), 73–101.

Mandelbrot, B. (1963). The variation of certain speculative prices. Journal of Business, 36(4), 392–417.

Markowitz, H. (1952). Portfolio selection. Journal of Finance, 7(1), 77–91.

McCullagh, P., & Nelder, J. A. (1989). Generalized Linear Models. Chapman and Hall.

Rousseeuw, P. J. (1984). Least median of squares regression. Journal of the American Statistical Association, 79(388), 871–880.

Sharpe, W. F. (1964). Capital asset prices: A theory of market equilibrium under conditions of risk. Journal of Finance, 19(3), 425–442.

Theil, H. (1950). A rank-invariant method of linear and polynomial regression analysis. Proceedings of the Koninklijke Nederlandse Akademie van Wetenschappen, 53, 386–392.

White, H. (1980). A heteroskedasticity-consistent covariance matrix estimator and a direct test for heteroskedasticity. Econometrica, 48(4).

Comparing Stock
Risk Using F-Test

9.1 Introduction

A F-test is a method of inferential statistics that determines the statistical difference between the variances of two variables. The F-test is applied when we want to compare and check whether the variances of two samples are equal or not. We apply the F-test on data that is normally distributed and samples that are independent variables. The F-test is based on F-distribution. The rejection zone is decided by analyzing and comparing critical values. Here, the F-test is performed to compare the risk of two stocks.

The degree of freedom represents the number of observations that are considered for the calculation of the chi-square variables used for calculating the ratio. As the degree of freedom increases, the F-distribution becomes more symmetrical and approaches the bell-shaped normal distribution.

9.1.1 Review of Literature

F-test is applied for comparing the variances of two variables (Montgomery et al., 2017; Allen 2012). The SciPy library is a tool for performing F-test (Vertanen, 2020; Montgomery et al., 2017; Pedregosa et al., 2011; Seabold & Perktold, 2010). F-test is applied for financial analysis of stock markets (Zou et al., 2020; Virtanen et al., 2020). F-test is used for feature engineering on high dimensional data, and F-test is integrated with machine learning models and the workflow for predictive analytics is prepared (Pedregosa et al., 2011). Johnson and Wichern (2007) conducted a conceptual research on its application and constraints. The statsmodels library provides the workflow documentation for applying F-test (Seabold & Perktold, 2010; Gelman et al., 2013). F-test should be applied after checking the normality of the data.

DOI: 10.1201/9781032618241-9

9.2 Research Methodology

9.2.1 Data Source

The Yahoo Finance financial database is used to perform the F-test.

9.2.2 Period of Study

The study period was from 1 January 2023 to 12 January 2023 (Refer Figure 9.2). The interval for selected data is the daily closing stock price of two companies for analysis. The study used a sample size of 12 days.

9.2.3 Software Used for Data Analysis

Python Programming, Anaconda (Refer Figure 9.1)

9.2.4 Model Applied

For this study, we applied the F-test.

9.2.5 Limitations of the Study

The study is restricted to t-tests only.

9.2.6 Future Scope of the Study

In the future, the study can be conducted for different stocks at the same time.

Hypothesis

Null Hypothesis—The variance of group 1 stock is equal to the variance of group 2 stock (same risk).

Alternative Hypothesis—The variance of group 1 stock is not equal to the variance of group 2 stock.

```
In [1]:  import numpy as np
         import scipy.stats as stats
         import pandas as pd
         import numpy as np
```

Figure 9.1 Python libraries for performing an F-test.

```
In [2]: data = pd.read_csv (r'C:\Users\nitin\Desktop\F.csv')
        print (data)
```

```
        Date     group1  group2
0    1/1/2023  18.799999   4.67
1    2/1/2023  17.940001   4.63
2    3/1/2023  17.440001   4.70
3    4/1/2023  15.540000   4.88
4    5/1/2023  15.960000   4.72
5    6/1/2023  16.070000   4.89
6    7/1/2023  16.660000   4.89
7    8/1/2023  17.370001   4.84
8    9/1/2023  17.110001   4.55
9   10/1/2023  16.420000   4.85
10  11/1/2023  17.549999   4.85
11  12/1/2023  18.379999   5.57
```

Figure 9.2   Fetching the data sets into a Python environment for performing an F-test.

```
In [3]:  # Create the data for two groups
         group1 = np.random.rand(11)
         group2 = np.random.rand(11)

         # Calculate the sample variances
         variance1 = np.var(group1, ddof=1)
         variance2 = np.var(group2, ddof=1)

         # Calculate the F-statistic
         f_value = variance1 / variance2

         # Calculate the degrees of freedom
         df1 = len(group1) - 1
         df2 = len(group2) - 1

         # Calculate the p-value
         p_value = stats.f.cdf(f_value, df1, df2)

         # Print the results
         print('Degree of freedom 1:',df1)
         print('Degree of freedom 2:',df2)
         print("F-statistic:", f_value)
         print("p-value:", p_value)

         Degree of freedom 1: 10
         Degree of freedom 2: 10
         F-statistic: 0.4831972166614582
         p-value: 0.1335063153601323
```

Figure 9.3   Performing an F-test in Python using **scipy**.stats library.

Conclusion the p-value of the test is 0.13, which is more than the alpha value of 0.05 (Refer Figure 9.3). Hence, we cannot reject the null hypothesis of the test. Based on the above analysis, we conclude that the variance of return of both the stocks is not different, thus rejecting the alternative hypothesis.

References

Allen, F., & Powell, M. (2012). Market Liquidity: A Primer. Oxford University Press.

Gelman, A., Carlin, J. B., Stern, H. S., Dunson, D. B., Vehtari, A., & Rubin, D. B. (2013). Bayesian Data Analysis (Vol. 2). CRC Press.

Johnson, R. A., & Wichern, D. W. (2007). Applied Multivariate Statistical Analysis. Prentice Hall.

Montgomery, D. C., Jennings, C. L., & Kulahci, M. (2017). Introduction to Time Series Analysis and Forecasting. John Wiley & Sons.

Pedregosa, F., Varoquaux, G., Gramfort, A., Michel, V., Thirion, B., Grisel, O., . . . & Vanderplas, J. (2011). Scikit-learn: Machine learning in Python. Journal of Machine Learning Research, 12(Oct), 2825–2830.

Seabold, S., & Perktold, J. (2010). Statsmodels: Econometric and statistical modeling with Python. Proceedings of the 9th Python in Science Conference (pp. 92–96).

Virtanen, P., Gommers, R., Oliphant, T. E., Haberland, M., Reddy, T., Cournapeau, D., . . . & van der Walt, S. J. (2020). SciPy 1.0: Fundamental algorithms for scientific computing in Python. Nature Methods, 17(3), 261–272.

Zou, H., Hastie, T., & Tibshirani, R. (2020). Sparse principal component analysis. Journal of Computational and Graphical Statistics, 27(2), 316–324.

10

STOCK RISK ANALYSIS USING T-TEST

10.1 Introduction

A t-test is a method of inferential statistics that determines the statistical difference between the means of two variables (Lumley et al., 2002). The t-test that was developed by Gosset (1908) is an important tool in statistics for comparing the means of two variables which are significantly different. The t-test is applied in healthcare, social sciences, etc. (Field, 2018; Howell, 2013; Tabachnick 2019). The t-test is applied in different forms such as one-tailed and two-tailed paired test for different experiments in research (Cohen, 1988; Hinkle et al., 2003; Mendenhall & Sincich, 2016). The validity of t-test depends upon the assumption that the data is normally distributed and is independent (Urdan, 2016). When the above assumptions are violated, we applied the U-test (McDonald, 2014). Despite lots of limitations, we still apply t-test because of its simplicity (Sheskin, 2003; Tabachnick & Fidell, 2019). The present study on t-tests will further modify its application and improve its accuracy (Neter et al., 1996; Rosenthal & Rosnow, 2008). It is considered to be the basic test for hypothesis testing (Trochim et al., 2016; Zar, 2010; Mendenhall & Sincich, 2016).

10.2 Research Methodology

10.2.1 Data Source

The Yahoo Finance financial database is used to perform a t-test.

10.2.2 Period of Study

The study period was from 1 January 2023 to 12 January 2023 (Refer Figure 10.2). The interval for selected data is the daily closing stock price of two companies for analysis. The study used a sample size of 12 days.

DOI: 10.1201/9781032618241-10

10.2.3 Software Used for Data Analysis

Python Programming, Anaconda (Refer Figure 10.1)

10.2.4 Model Applied

For this study, we applied the t-test.

10.2.5 Limitations of the Study

The study is restricted to t-tests only.

10.2.6 Future Scope of the Study

In the future, the study can be conducted for different stocks at the same time.

Hypothesis

> **Null Hypothesis**—The mean of group 1 stock is equal to the mean of group 2 stock (same risk).
> **Alternative Hypothesis**—The mean of group 1 stock is not equal to the mean of group 2 stock.

The p-value of the test is 0.23 (Refer Figure 10.3), which is more than the alpha value of 0.05. Hence, we cannot reject the null hypothesis of the test. From the above analysis, we conclude that the mean return of both stocks is different, thus rejecting the alternative hypothesis (Refer Figure 10.4).

```
In [1]:  # Python program to demonstrate how to
         # perform two sample T-test

         # Import the library
         import scipy.stats as stats
         import numpy as np
         import scipy.stats as stats
         import pandas as pd
         import numpy as np
```

Figure 10.1   Python libraries for performing a t-test.

```
In [2]:  data = pd.read_csv (r'C:\Users\nitin\Desktop\F.csv')
         print (data)

               Date     group1   group2
         0    1/1/2023  18.799999   4.67
         1    2/1/2023  17.940001   4.63
         2    3/1/2023  17.440001   4.70
         3    4/1/2023  15.540000   4.88
         4    5/1/2023  15.960000   4.72
         5    6/1/2023  16.070000   4.89
         6    7/1/2023  16.660000   4.89
         7    8/1/2023  17.370001   4.84
         8    9/1/2023  17.110001   4.55
         9   10/1/2023  16.420000   4.85
         10  11/1/2023  17.549999   4.85
         11  12/1/2023  18.379999   5.57
```

Figure 10.2   Fetching the data sets into a Python environment for performing a t-test.

```
In [5]:  # Print the variance of both data groups
         print(np.var(group1), np.var(group2))

         0.12303191538629704 0.12295852542416214
```

Figure 10.3   Calculating the variance.

```
In [4]:  # Create the data for two groups
         group1 = np.random.rand(11)
         group2 = np.random.rand(11)
         # Perform the two sample t-test with equal variances
         stats.ttest_ind(group1, group2, equal_var=True)

Out[4]:  Ttest_indResult(statistic=1.2193142946867186, pvalue=0.2369105013488233)
```

Figure 10.4   Performing a t-test in Python using scipy.stats library.

10.3 Conclusion

The t-test is performed on a small sample size and the average return of two stock variables are calculated by applying Python libraries like scipy.stats. The study period was from 1 January 2023 to 12 January 2023. The interval for selected data is the daily closing stock price of two companies for analysis. The p-value of the t-test is 0.23, which is more than the alpha value of 0.05. Hence, we cannot reject the null hypothesis of the test. From the above analysis, we conclude that the mean return of both stocks is different, which ultimately results in the rejection of the alternative hypothesis.

References

Box Cohen, J. (1988). Statistical Power Analysis for the Behavioral Sciences (2nd ed.). Erlbaum.

Field, A. (2018). Discovering Statistics using IBM SPSS Statistics (5th ed.). Sage.

Gosset, W. S. (1908). The probable error of a mean. Biometrika, 6(1), 1–25.

Hinkle, D. E., Wiersma, W., & Jurs, S. G. (2003). Applied Statistics for the Behavioral Sciences (5th ed.). Houghton Mifflin.

Howell, D. C. (2013). Statistical Methods for Psychology (8th ed.). Cengage Learning.

Lumley, T., Diehr, P., Emerson, S., & Chen, L. (2002). The importance of the normality assumption in large public health data sets. Annual Review of Public Health, 23, 151–169.

McDonald, J. H. (2014). Handbook of Biological Statistics (3rd ed.). Sparky House Publishing.

Mendenhall, W., & Sincich, T. (2016). Statistics for Engineering and the Sciences (6th ed.). Pearson.

Neter, J., Kutner, M. H., Nachtsheim, C. J., & Wasserman, W. (1996). Applied Linear Statistical Models (4th ed.). McGraw-Hill.

Rosenthal, R., & Rosnow, R. L. (2008). Essentials of Behavioral Research: Methods and Data Analysis (3rd ed.). McGraw-Hill.

Sheskin, D. J. (2003). Handbook of Parametric and Nonparametric Statistical Procedures (3rd ed.). CRC Press.

Tabachnick, B. G., & Fidell, L. S. (2019). Using Multivariate Statistics (7th ed.). Pearson.

Trochim, W. M. K., Donnelly, J. P., & Arora, K. (2016). Research Methods: The Essential Knowledge Base (2nd ed.). Cengage Learning.

Urdan, T. C. (2016). Statistics in Plain English (4th ed.). Routledge.

Zar, J. H. (2010). Biostatistical Analysis (5th ed.). Pearson Prentice Hall.

STOCK INVESTMENT STRATEGY USING A Z-SCORE

11.1 Introduction to Z-Score

Z-score represents or gives us an interpretation of the number of standard deviations the value of a variable is from the average or mean. It is measured in terms of standard deviation from its average or mean. Z-score value of 2.0 indicates that it is two time the standard deviation away from the mean. A positive Z-score indicates that the value is above the mean and a negative Z-score indicates that the value is below the mean. If the Z-score is zero, then it is equal to the mean. The Z-score is represented by the below formula:

$$Z \text{ score} = \frac{\text{Value}(X) - \text{Mean}(\mu)}{\text{Standard Deviation}(\sigma)}$$

where

X is the variable.

μ is the mean, which is given by the value of the variable divided by the number of items.

σ is the standard deviation of variable X.

The Z-score model is developed and implemented with its evaluation in five steps:

1. Introduction to the Z-score
2. Fetching the data into a Python environment
3. Calculating the Z-score for the stock
4. Result analysis by evaluating the Z-score
5. Conclusion

Aanderson (1962) before applying the Z-test we need to test the normality of the data. The normality of data is considered to be a void

 DOI: 10.1201/9781032618241-11

assumption as far as stock market data is considered (Mandelbrot, 1963; Fama, 1965; Campbell et al., 1997; Fama, 1965; Hampel, 1974; Huber, 1964; Jensen, 1969; Greene 2003). The significance of return is tested by applying Z-testfor stock market analysis. The Z-test is applied to optimize the return on portfolio (Markowitz, 1952; Rousseeuw, 1984; Theil, 1950; White, 1980). Because the Z-test does not accurately capture the volatility of the market, non-parametric tests can be used as an alternative method. The assumption that the Z-test considers equal variance (Brown & Forsythe, 1974) cannot be held in the case of stock market data. Hence, it is important to check homoscedasticity before the application of Z-test (Aanderson, 1962; Brown & Forsythe, 1974).

11.2 Applied Research Methodology

11.2.1 Data Source

Data is taken from Yahoo Finance, which is a reliable source.

11.2.2 Sample Size

The daily price of the MRF stock is considered for the study from 2 January 2023 to 27 February 2024 (daily stock price) (Refer Figure 11.1).

11.2.3 Software Used for Data Analysis

Python Programming libraries used for analysis are Pandas, *NumPy*, and *SciPy*.

11.2.4 Model Applied

The Z-score model is applied for analysis.

11.3 Fetching the Data into a Python Environment and Defining the Dependent and Independent Variables

Raw data filtering is a procedure applied in feature engineering (Refer Figure 11.1). Feature engineering is a process of converting raw data into features that can be utilized in an ML model. The data was fetched in the Python Anaconda environment using the Jupyter Notebook, as the format of the data file was not readable in Python. A data frame is created by fetching the comma-separated values (CSV) file, making it readable in Python and utilizing it for further processing in the form of a data frame. The data frame is created as per the model requirement. The data frame needs to be structured in as per the requirement of the model. The first step in creating a data frame is to structure the data so that the program can read and work on the data. Once the data frame is created, it is ready to be used by the algorithm. The syntax used for creating a data frame in Python Programming is presented in Figure 11.1.

```
In [1]:  import pandas as pd
         import numpy as np
         import scipy.stats as stats

In [2]:  data = pd.read_csv (r'C:\Users\nitin\Desktop\REG.csv')
         print (data)
```

	Open	High	Low	Close	Adj Close	Volume
0	18131.69922	18215.15039	18086.50000	18197.44922	18197.44922	256100
1	18163.19922	18251.94922	18149.80078	18232.55078	18232.55078	208700
2	18230.65039	18243.00000	18020.59961	18042.94922	18042.94922	235200
3	18101.94922	18120.30078	17892.59961	17992.15039	17992.15039	269000
4	18008.05078	18047.40039	17795.55078	17859.44922	17859.44922	238200
5	17952.55078	18141.40039	17936.15039	18101.19922	18101.19922	257200
6	18121.30078	18127.59961	17856.00000	17914.15039	17914.15039	283300
7	17924.25000	17976.34961	17824.34961	17895.69922	17895.69922	259900
8	17920.84961	17945.80078	17761.65039	17858.19922	17858.19922	227800
9	17867.50000	17999.34961	17774.25000	17956.59961	17956.59961	256700
10	18033.15039	18049.65039	17853.65039	17894.84961	17894.84961	206200
11	17922.80078	18072.05078	17886.94922	18053.30078	18053.30078	219100
12	18074.30078	18183.75000	18032.44922	18165.34961	18165.34961	258800
13	18119.80078	18155.19922	18063.75000	18107.84961	18107.84961	237800
14	18115.59961	18145.44922	18016.19922	18027.65039	18027.65039	237200
15	18118.44922	18162.59961	18063.44922	18118.55078	18118.55078	202500
16	18183.94922	18201.25000	18078.65039	18118.30078	18118.30078	216900
17	18093.34961	18100.59961	17846.15039	17891.94922	17891.94922	257200
18	17877.19922	17884.75000	17493.55078	17604.34961	17604.34961	476300
19	17541.94922	17709.15039	17405.55078	17648.94922	17648.94922	432400
20	17731.44922	17735.69922	17537.55078	17662.15039	17662.15039	398300
21	17811.59961	17972.19922	17735.40039	17616.30078	17616.30078	512900
22	17517.09961	17653.90039	17445.94922	17610.40039	17610.40039	490100
23	17721.75000	17870.30078	17584.19922	17854.05078	17854.05078	424100
24	17818.55078	17823.69922	17698.34961	17764.59961	17764.59961	282500
25	17790.09961	17811.15039	17652.55078	17721.50000	17721.50000	354400
26	17750.30078	17898.69922	17744.15039	17871.69922	17871.69922	291000
27	17885.50000	17916.90039	17779.80078	17893.44922	17893.44922	266900
28	17847.55078	17876.94922	17801.00000	17856.50000	17856.50000	232000
29	17859.09961	17880.69922	17719.75000	17770.90039	17770.90039	231300
...	...	...	...	...	...	...
255	22053.15039	22115.55078	21963.55078	22097.44922	22097.44922	345500
256	22080.50000	22324.15039	21969.80078	22032.30078	22032.30078	292400
257	21647.25000	21851.50000	21550.44922	21571.94922	21571.94922	456000
258	21414.19922	21530.40039	21285.55078	21462.25000	21462.25000	387300
259	21615.19922	21670.50961	21575.00000	21622.40039	21622.40039	343100
260	21716.69922	21750.25000	21192.59961	21238.80078	21238.80078	469700
261	21185.25000	21482.34961	21137.19922	21453.94922	21453.94922	407500
262	21454.59961	21459.00000	21247.05078	21352.59961	21352.59961	418100
263	21433.09961	21763.25000	21429.59961	21737.59961	21737.59961	376700
264	21775.75000	21813.05078	21501.80078	21522.09961	21522.09961	375100
265	21487.25000	21741.34961	21448.84961	21725.69922	21725.69922	410600
266	21780.65039	21832.94922	21658.75000	21697.44922	21697.44922	332500
267	21812.75000	22126.80078	21805.55078	21853.80078	21853.80078	442800
268	21921.05078	21964.30078	21726.94922	21771.69922	21771.69922	440800
269	21825.19922	21951.40039	21737.55078	21929.40039	21929.40039	371000
270	22045.05078	22053.30078	21860.15039	21930.50000	21930.50000	346300
271	22009.65039	22011.05078	21665.30078	21717.94922	21717.94922	491100
272	21727.00000	21804.44922	21629.90039	21782.50000	21782.50000	349200
273	21800.80078	21831.69922	21574.75000	21616.05078	21616.05078	287400
274	21578.59961	21766.80078	21543.34961	21743.25000	21743.25000	365800
275	21578.15039	21870.84961	21530.19922	21840.05078	21840.05078	359100
276	21906.55078	21953.84961	21794.80078	21910.75000	21910.75000	345400
277	22020.30078	22068.65039	21968.94922	22040.69922	22040.69922	343300
278	22103.44922	22186.65039	22021.05078	22122.25000	22122.25000	0
279	22099.19922	22215.59961	22045.84961	22196.94922	22196.94922	295700
280	22248.84961	22249.40039	21997.94922	22055.05078	22055.05078	364500
281	22081.15039	22252.50000	21875.25000	22217.44922	22217.44922	343500
282	22290.00000	22297.50000	22186.09961	22212.69922	22212.69922	226000
283	22169.19922	22202.15039	22075.15039	22122.05078	22122.05078	207800
284	22090.19922	22218.25000	22085.65039	22198.34961	22198.34961	252200

[285 rows x 6 columns]

Figure 11.1   Creating a data frame.

11.4 Calculating the Z-Score for the Stock

The Z-score analysis for the stock is done by calculating the Z-score for opening price, closing price day high, day low, and stock volume (Refer Figure 11.2). Calculating the Z-score and determining risk is done by comparing the values of stock with parameters like opening price, closing price day high, day low, and stock volume. The value of a stock is positive and above zero means that it is above the mean and if it is below zero it is considered to be negative or below the mean. The Z-score of zero indicates the value is equal to zero. If the value is above it is a good investment opportunity and if the Z-score is below zero it means the value is going down.

```
In [3]:  data.apply(stats.zscore)
```

Out[3]:

	Open	High	Low	Close	Adj Close	Volume
0	-0.827672	-0.811017	-0.793615	-0.771011	-0.771011	-0.196261
1	-0.805640	-0.785432	-0.749194	-0.746576	-0.746576	-0.776847
2	-0.758461	-0.791654	-0.839656	-0.878661	-0.878661	-0.451699
3	-0.848481	-0.876963	-0.929275	-0.913924	-0.913924	-0.026939
4	-0.914157	-0.927649	-0.997225	-1.006300	-1.006300	-0.414889
5	-0.952976	-0.862293	-0.898783	-0.838012	-0.838012	-0.181764
6	-0.834945	-0.871889	-0.954901	-0.968221	-0.968221	0.138476
7	-0.972771	-0.977048	-0.977061	-0.981065	-0.981065	-0.148636
8	-0.975149	-0.998287	-1.020960	-1.007170	-1.007170	-0.542495
9	-1.012464	-0.961057	-1.012139	-0.938671	-0.938671	-0.187899
10	-0.896601	-0.926084	-0.956546	-0.981657	-0.981657	-0.807621
11	-0.973786	-0.910510	-0.933231	-0.871366	-0.871366	-0.649241
12	-0.867819	-0.832849	-0.831359	-0.793356	-0.793356	-0.198942
13	-0.835994	-0.852700	-0.809443	-0.833383	-0.833383	-0.419797
14	-0.838933	-0.859478	-0.842736	-0.889211	-0.889211	-0.427159
15	-0.836940	-0.847664	-0.809654	-0.825934	-0.825934	-0.852919
16	-0.791126	-0.820682	-0.799010	-0.826108	-0.826108	-0.676235
17	-0.864495	-0.890661	-0.961797	-0.983676	-0.983676	-0.181764
18	-1.006680	-1.040734	-1.208672	-1.183880	-1.183880	2.606635
19	-1.240169	-1.162823	-1.270286	-1.162833	-1.162833	1.967894
20	-1.107624	-1.144364	-1.177865	-1.143644	-1.143644	1.549496
21	-1.061564	-0.979933	-1.306799	-1.175561	-1.175561	2.956607
22	-1.257649	-1.201236	-1.242001	-1.179668	-1.179668	2.675857
23	-1.114408	-1.060780	-1.145204	-1.010058	-1.010058	1.866056
24	-1.046702	-1.083181	-1.065281	-1.072327	-1.072327	0.128660
25	-1.066602	-1.091905	-1.097347	-1.102329	-1.102329	1.010864
26	-1.094439	-1.031036	-1.033213	-0.997772	-0.997772	0.232963
27	-0.999874	-1.018381	-1.008252	-0.982632	-0.982632	-0.136366
28	-1.026418	-1.046158	-0.993409	-1.008363	-1.008363	-0.490962
29	-1.018340	-1.043650	-1.060297	-1.067940	-1.067940	-0.499651
...	...	...	...	...	...	...
255	1.916159	1.900804	1.921030	1.943862	1.943862	0.901653
256	1.934289	1.906783	1.925406	1.898511	1.898511	0.260130
257	1.631255	1.717218	1.631794	1.578061	1.578061	2.257469
258	1.468249	1.600226	1.446324	1.501687	1.501687	1.414528
259	1.608837	1.591444	1.648983	1.613171	1.613171	0.872206
260	1.679831	1.646822	1.381243	1.346139	1.346139	2.180160
261	1.308112	1.460560	1.342454	1.495908	1.495908	1.662377
262	1.496607	1.444326	1.419367	1.425357	1.425357	1.792436
263	1.481469	1.655861	1.547180	1.693363	1.693363	1.284469
264	1.721133	1.690486	1.597732	1.643349	1.643349	1.264838
265	1.519344	1.640634	1.560658	1.685079	1.685079	1.700413
266	1.724661	1.704321	1.707622	1.665414	1.665414	0.742147
267	1.747013	1.908826	1.810406	1.774254	1.774254	2.096499
268	1.822763	1.795645	1.766372	1.717101	1.717101	2.070959
269	1.766720	1.786676	1.762795	1.826880	1.826880	1.214632
270	1.909494	1.857524	1.848634	1.827646	1.827646	0.911469
271	1.884733	1.828149	1.712208	1.679684	1.679684	2.688127
272	1.687036	1.684605	1.687422	1.724620	1.724620	0.947052
273	1.738655	1.703461	1.648808	1.608751	1.608751	0.188782
274	1.643181	1.658330	1.626823	1.697297	1.697297	1.150729
275	1.582924	1.730671	1.617616	1.764682	1.764682	1.068622
276	1.812621	1.788379	1.802879	1.813897	1.813897	0.900427

Figure 11.2 Calculating the Z-score for the stock.

11.5 Results Z-Score Analysis

The mean of the opening price is 7.86 of the Z-score (Refer Figure 11.3). So we can interpret this as it is the best time to invest since the opening price has given a good return and the Z-score is above seven times of mean return. The Z-score value of the variables High, Low, and Close has given a lower return since the standard deviation of one. The Z-score of minimum ranges from –1.63 to –1.61 which indicates low risk and high return at the opening price as the Z-score mean is high, The maximum Z-score ranges from 2.08 for the opening price and 2.02 for all other variables which show high return and high risk in the opening price.

```
In [5]:  data.apply(stats.zscore).describe()
Out[5]:
```

	Open	High	Low	Close	Adj Close	Volume
count	2.850000e+02	2.850000e+02	2.850000e+02	2.850000e+02	2.850000e+02	2.850000e+02
mean	7.868949e-17	2.031903e-15	-1.160865e-15	-2.017879e-15	-2.017879e-15	-3.885781e-16
std	1.001759e+00	1.001759e+00	1.001759e+00	1.001759e+00	1.001759e+00	1.001759e+00
min	-1.635108e+00	-1.612939e+00	-1.674417e+00	-1.642832e+00	-1.642832e+00	-3.337541e+00
25%	-8.484805e-01	-8.622934e-01	-8.427361e-01	-8.632115e-01	-8.632115e-01	-5.633531e-01
50%	4.942933e-02	4.301416e-02	6.249725e-02	5.518194e-02	5.518194e-02	-1.952610e-01
75%	3.694245e-01	3.516437e-01	3.891561e-01	3.676357e-01	3.676357e-01	3.041173e-01
max	2.080822e+00	2.027308e+00	2.076849e+00	2.027397e+00	2.027397e+00	5.208331e+00

Figure 11.3   Z-scores with descriptive statistics for risk analysis.

11.6 Conclusion

The four continuous variables are Open, Close, High, and Low which have poor Z-scores. Only the open stock price is exceptional and has a higher return and high risk since the mean of Z-score is 7.86 and the standard deviation is low which indicates that the best investment opportunity is the Opening price.

References

Aanderson, T. W. (1962). An Introduction to Multivariate Statistical Analysis. Wiley.

Brown, M. B., & Forsythe, A. B. (1974). Robust tests for equality of variances. Journal of the American Statistical Association, 69(346), 364–367.

Campbell, J. Y., Lo, A. W., & MacKinlay, A. C. (1997). The Econometrics of Financial Markets. Princeton University Press.

Fama, E. F. (1965). The behavior of stock prices. Journal of Business, 38(1), 34–105.

Greene, W. H. (2003). Econometric Analysis. Prentice Hall.

Hampel, F. R. (1974). The influence curve and its role in robust estimation. Journal of the American Statistical Association, 69(346), 383–393.

Huber, P. J. (1964). Robust estimation of a location parameter. Annals of Mathematical Statistics, 35(1), 73–101.

Jensen, M. C. (1969). Risk, the pricing of capital assets, and the evaluation of investment portfolios. Journal of Business, 42(2), 167–247.

Mandelbrot, B. (1963). The variation of certain speculative prices. Journal of Business, 36(4), 392–417.

Markowitz, H. (1952). Portfolio selection. Journal of Finance, 7(1), 77–91.

Rousseeuw, P. J. (1984). Least median of squares regression. Journal of the American Statistical Association, 79(388), 871–880.

Sharpe, W. F. (1964). Capital asset prices: A theory of market equilibrium under conditions of risk. Journal of Finance, 19(3), 425–442.

Theil, H. (1950). A rank-invariant method of linear and polynomial regression analysis. Proceedings of the Koninklijke Nederlandse Akademie van Wetenschappen, 53, 386–392.

White, H. (1980). A heteroskedasticity-consistent covariance matrix estimator and a direct test for heteroskedasticity. Econometrica, 48(4), 817–838.

12

APPLYING A SUPPORT VECTOR MACHINE MODEL USING PYTHON PROGRAMMING

12.1 Introduction

Supervised machine learning models include the support vector machine (SVM) as one of the reliable models in predictive analytics. The model was developed in 1960, and further its application and accuracy were effectively increased by the 1990s. The model had tremendous accuracy and was known for achieving good results with precise accuracy. It stands unique compared to other machine learning models as it has the minimum classification errors. As we closely analyze the machine learning models, we will learn about the accuracy of the SVM compared to other machine learning models (Refer Figure 12.2). The algorithm differentiates the data points with precise accuracy. The SVM algorithm can classify with tremendous accuracy and create a model that can give good results compared to other machine learning models.

The support vector machine algorithm is the most used and applied machine learning classification algorithm. The SVM algorithm is used since it outperforms other machine learning classification models like the logistic regression model and Naive Bayes model. The SVM algorithm gives more optimum solutions than any other machine learning classification model. The SVM algorithm is also known for the accuracy it provides (Refer Figure 12.2). The main utility of the SVM algorithm is to find out the hyperplane (N-dimensional) that creates a difference in data points (refer to Figure 12.1). To bifurcate the data as per classes, many hyperplanes can be drawn. The optimum hyperplane, which has the maximum margin, is considered to be the best (refer to Figure 12.1). The decision boundaries are hyperplanes that help classify the data points.

 DOI: 10.1201/9781032618241-12

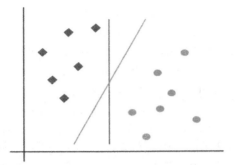

Figure 12.1   Different bifurcation of the data as per classes.

(Source: https://www.javatpoint.com)

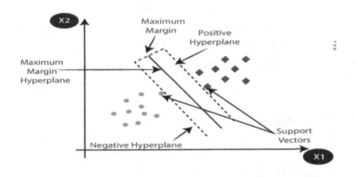

Figure 12.2   Different bifurcation of the data with maximum margin hyperplane.

(Source: https://www.javatpoint.com)

12.1.1 Review of Literature

In the area of machine learning, support vector machine is considered to be a powerful algorithm for classification and regression analysis (Burbidge et al., 2001). The support vector machine has a wide range of applications (Dhillon & Verma, 2020). Hence, a literature review is conducted in order to grab various related studies and their role in the development of support vector machines. The SVM is being (Varma) applied for image processing in order to improve the clarity and quality of images further by applying high-quality image processing, the image recognition system's is created.

Deo (2015) applied the support vector machine model in health analytics in order to identify complex medical patterns in predictive health analytics (Cervantes et al., 2023). Jardine et al. (2006) applied a support vector machine learning model in automation for manufacturing and predicted machine failure by analyzing the past data, which improves the maintenance cost. Garcia-Lamont et al. (2023) applied support vector machine learning model for structure safety in the area of infrastructure safety Hinton et al., 2012; Kim 2014; Nguyen et al., 2020; Schölkopf et al., 2001. Toledo-Pérez et al. (2019) applied the SVM algorithm and improved the signal classification accuracy. Huang et al. (2005) applied SVM models in stock market prediction in order to help investors in making investment decisions. Joachims (1998) applied SVM models for natural language processing NLP and improved the accuracy of a documents classification system. Ding and Dubchak (2001) applied SVM models for understanding and classification of the protein structure. Mountrakis et al. (2011) applied SVM models in the remote sensing field and achieved a high data classification accuracy (Burges, 1998).

12.2 Research Methodology

12.2.1 Data Collection

Secondary data was collected from Yahoo Finance.

12.2.2 Sample Size

Daily stock price of the MRF stock is considered for the study from 2/1/2023 to 5/1/2024.

12.2.3 Software Used for Data Analysis

Python Programming

12.2.4 Model Applied

For this study, we applied the support vector machine algorithm.

12.2.5 Limitations of the Study

The study is limited to only predicting the stock price of MRF.

12.2.6 Future Scope of the Study

In the future, the study can be extended to compare SVM models applied to different sectors of industry at the macro level.

12.3 Methodology

For creating a predictive model, we selected and applied the support vector machine algorithm.

Research is carried out in five steps:

12.4 Feature Engineering and Data Processing
12.5 Training and Testing
12.6 Predicting a Support Vector Machine Model with a Confusion Matrix
12.7 Calculating False Negative, False Positive, True Negative, and True Positive
12.8 Results and analysis

12.4 Feature Engineering and Data Processing

The process of converting raw data into features that can be easily utilized to create a model as per the requirement of the algorithm is called feature engineering (Refer Figure 12.3). The creation of a data frame is the first step in creating a model. The data frame is prepared to maintain the notion of the model which has different variables. The feature engineering is the process of preparing the data according to the need and required to be converted into nominal scale, ordinal scale, etc. Feature engineering prepares data that can be read and utilized by algorithms. Further, it converts raw data which will be ready for the program to utilize it best possible manner. The syntax used for creating a data frame in Python Programming is presented in Figure 12.3.

Figure 12.3   Creating a data frame.

12.5 Training and Testing

To conduct the study, secondary data was collected from Yahoo Finance (Refer Figure 12.4). The dependent variable for predicting (Y) is the binary class (Figure 12.4) and the four independent variables 'High', 'Low', 'Open', and 'Close' are continuous in nature.

VARIABLE	CLASSES
Buy/Sell (Dependent)	Tomorrow's Price > Today's Price Buy = 1 Tomorrow's Price < Today's Price Sell = 0
Open (Independent)	Continuous
Close (Independent)	Continuous
High (Independent)	Continuous
Low (Independent)	Continuous

```
In [3]:  # defining the independent and dependent variables

         # X is the independent variables
         X = ip_data[['High','Low','Open','Close']]

         # y is the dependent variable
         y = np.where(ip_data['Adj Close'].shift(1) > ip_data['Adj Close'], 0, 1)
```

Figure 12.4   Defining the dependent and independent variables.

For trial and testing, the data is divided into two categories (Refer Figure 12.5). 80 percent of data is converted and used for trial and 20 percent of data is used for testing to evaluate. With trial and testing, the test results are validated by creating a confusion matrix.

```
In [4]:  X_train, X_test, y_train, y_test = sklearn.model_selection.train_test_split(X, y, test_size = 0.20, random_state = 5)
         print(X_train.shape)
         print(X_test.shape)
         print(y_train.shape)
         print(y_test.shape)

         (200, 4)
         (50, 4)
         (200,)
         (50,)
```

Figure 12.5   The Python code for trial and testing.

12.6 Predicting a Support Vector Machine Model with a Confusion Matrix

12.6.1 Creating a Confusion Matrix

A confusion matrix measures model performance (Refer Table 12.1). It evaluates the actual values and predicted values. It is of the order of N X N, where N is the class of dependent/target variable. For binary classes, it is 2 X 2 confusion matrix.

12.7 Calculating False Negative, False Positive, True Negative, and True Positive

The confusion matrix for our data set is as follows:

Table 12.1 The Confusion Matrix

	0	1
0	22 (TP)	4 (FN)
1	0 (FP)	24 (TN)

```
In [5]: from sklearn.svm import SVC
        from sklearn.metrics import accuracy_score

        clf = SVC(kernel='linear')
        clf.fit(X_train,y_train)
        y_pred = clf.predict(X_test)
        print(accuracy_score(y_test,y_pred))

        0.92
```

```
In [6]: from sklearn.metrics import classification_report, confusion_matrix
        print(confusion_matrix(y_test,y_pred))
        print(classification_report(y_test,y_pred))
        Results

        [[22  4]
         [ 0 24]]
                       precision    recall  f1-score   support

                   0        1.00      0.85      0.92        26
                   1        0.86      1.00      0.92        24

        avg / total        0.93      0.92      0.92        50
```

Figure 12.6 The Python code for confusion matrix and classification report.

12.7.1 *Result Analysis*

12.7.1.1 *Accuracy Statistics*

It measures the overall accuracy of the model by analyzing the output predicted about incorrect predictions (Refer Figure 12.6).

To obtain the accuracy of the model, we apply the following formula:

Accuracy = True Positive + True Negative/True Positive + True Negative + False Positive + False Negative

$$\text{Accuracy} = \frac{22+4}{22+4+0+24} = 0.93$$

The accuracy for the overall model is 0.93

12.7.1.2 Recall

It is the ratio of true positive predictions divided by the total number of true positive predictions and false negative predictions. Higher recall implies more correct predictions (a small number of false negatives).

$$\text{Recall} = \frac{\text{True Positive}}{\text{True Positive} + \text{False Negative}}$$

$$\text{Recall} = \frac{22}{22 + 4} = 0.85$$

Recall for the overall model is 0.85

12.7.1.3 Precision

Precision measures how correctly we have predicted the true positives. It is the qualitative analysis of correctly predicted values.

$$\text{Precision} = \frac{\text{True Positive}}{\text{True Positive} + \text{False Positive}} = \frac{22}{22 + 0} = 1.00$$

The precision for the overall model is 1.00

12.8 Conclusion

The support vector machine model predicted the MRF stock with a precision of 100 percent. The overall model accuracy is 93 percent.

References

Burbidge, R., Trotter, M., Buxton, B., & Holden, S. (2001). Drug design by machine learning: Support vector machines for pharmaceutical data analysis. Computers & Chemistry, 26(1), 5–14. https://doi.org/10.1016/S0097-8485(01)00094-8

Burges, C. J. (1998). A tutorial on support vector machines for pattern recognition. Data Mining and Knowledge Discovery, 2(2), 121–167. https://doi.org/10.1023/A:1009715923555

Cervantes, J., Garcia-Lamont, F., Rodríguez-Mazahua, L., & López, A. (2023). A comprehensive survey on support vector machine classification: Applications, challenges and trends. Journal of Building Engineering. https://doi.org/10.1016/j.jobe.2023.104911

Deo, R. C. (2015). Machine learning in medicine. Circulation, 132(20), 1920–1930. https://doi.org/10.1161/CIRCULATIONAHA.115.001593

Dhillon, A., & Verma, G. K. (2020). Convolutional neural network: A review of models, methodologies and applications to object detection. Progress in Artificial Intelligence, 9(2), 85–112. https://doi.org/10.1007/s13748-019-00203-0

Ding, C., & Dubchak, I. (2001). Multi-class protein fold recognition using support vector machines and neural networks. Bioinformatics, 17(4), 349–358. https://doi.org/10.1093/bioinformatics/17.4.349

Garcia-Lamont, F., Cervantes, J., Rodríguez-Mazahua, L., & López, A. (2023). Support vector machine in structural reliability analysis: A review. Structural Safety. https://doi.org/10.1016/j.strusafe.2023.102211

Hinton, G., Deng, L., Yu, D., Dahl, G. E., Mohamed, A. r., Jaitly, N., . . . & Sainath, T. N. (2012). Deep neural networks for acoustic modeling in speech recognition: The shared views of four research groups. IEEE Signal Processing Magazine, 29(6), 82–97. https://doi.org/10.1109/MSP.2012.2205597

Huang, W., Nakamori, Y., & Wang, S. Y. (2005). Forecasting stock market movement direction with support vector machine. Computers & Operations Research, 32(10), 2513–2522. https://doi.org/10.1016/j.cor.2004.03.016

Jardine, Andrew & Lin, Daming & Banjevic, Dragan. (2006). A review on machinery diagnostics and prognostics implementing condition-based maintenance. Mechanical Systems and Signal Processing. 20. 1483–1510. 10.1016/j.ymssp.2005.09.012.

Joachims, T. (1998). Text categorization with support vector machines: Learning with many relevant features. European Conference on Machine Learning (pp. 137–142). https://doi.org/10.1007/BFb0026683

Kim, Y. (2014). Convolutional neural networks for sentence classification. EMNLP 2014. https://doi.org/10.3115/v1/D14-1181

Mountrakis, G., Im, J., & Ogole, C. (2011). Support vector machines in remote sensing: A review. ISPRS Journal of Photogrammetry and Remote Sensing, 66(3), 247–259. https://doi.org/10.1016/j.isprsjprs.2010.11.001

Nguyen, H. Q., Nguyen, N. D., & Nahavandi, S. (2020). A review on deep reinforcement learning.

Schölkopf, B., Platt, J. C., Shawe-Taylor, J., Smola, A. J., & Williamson, R. C. (2001). Estimating the support of a high-dimensional distribution. Neural Computation, 13(7), 1443–1471. https://doi.org/10.1162/089976601750264965

Toledo-Pérez, D. C., Rodríguez-Reséndiz, J., Gómez-Loenzo, R. A., & Jauregui-Correa, J. C. (2019). Support vector machine-based EMG signal classification techniques: A review. Applied Sciences, 9(20), 4402. https://doi.org/10.3390/app9204402

13

DATA VISUALIZATION FOR STOCK RISK COMPARISON AND ANALYSIS

13.1 Introduction to Data Visualization

Today due to the development of information technology and e-commerce in the world, data is being generated on an hourly basis. Financial data like the buying of stock and the movement of the market can be utilized by data visualization. The financial data contains certain trends and patterns which are very difficult to understand as the raw data is in raw format. Hence to overcome this problem, an attempt has been made by means of this study titled Data Visualization for Stock Risk Comparison and Analysis.

It is easy to analyze, observe, understand, and interpret data by visualization in Python Programming. Large amount of financial data can be analyzed and investment strategies can be made based on data visualization. The Python libraries which are mostly used are as follows:

1. Matplotlib
2. Seaborn
3. Bokeh
4. Plotly

Here we apply different Python libraries to create a scatter plot, line chart, bar chart, histogram, and bokeh.

13.1.1 Review of Past Studies

Visualizing data in Python is very easy and simple, which is the only reason why it has gained significant attraction and popularity (Smith, 2018; Hunter, 2007; Wang & Liu, 2020; Waskom et al.,

DOI: 10.1201/9781032618241-13

2020; Wickham & Grolemund, 2017). Python libraries, like matplot-lib, have been used due to their tremendous applications (Waskom et al., 2020). The existence of plotly and seaborn made its application in various areas. Jones et al. (2019) used Python for data analysis in the emerging areas of biology. Wang and Liu (2020) had uti-lized data visualization for financial data analytics and stock mar-ket analysis. McKinney (2017) applied a hybrid model in Python using Pandas (VanderPlas, 2016; Virtanen et al., 2020). Grolemund (2017) compared the utilization of data visualization by comparing R programming and Python. Python's extensive documentation and its user-interactive, user-friendly, and flexible nature have made data visualization with Python an important tool for researchers and aca-demicians (Hunter, 2007; Jones et al., 2019; McKinney, 2017; Smith, 2018).

13.1.2 *Applied Research Methodology*

13.1.2.1 *Data Source*
Data is taken from Yahoo Finance, which is a reliable source.

13.1.2.2 *Sample Size*
The daily price of the MRF stock is considered for the study from 2 January 2023 to 5 January 2024 (daily stock price).

13.1.2.3 *Software Used for Data Analysis*
Python Programming libraries used for analysis are statsmodels.api, Pandas, NumPy, and SciPy.

13.2 Fetching the Data into a Python Environment and Defining the Dependent and Independent Variables

Raw data filtering is a procedure applied in feature engineering (Refer Figure 13.1). Feature engineering is a process of convert-ing raw data into features that can be utilized in an ML model. The data was fetched in the Python Anaconda environment using the Jupyter Notebook as the format of the data file was not read-able in Python. A data frame is created by fetching the comma-separated values (CSV) file, making it readable in Python and utilizing it for further processing in the form of a data frame. The data frame is created as per the model requirement. The data frame needs to be structured in C as per the requirement of the model. The first step in creating a data frame is to structure the data so that the program can read and work on the data. Once the data frame is created, it is ready to be used by the algorithm. The syntax used for creating a data frame in Python Programming is presented in Figure 13.2.

```
In [1]:  import pandas as pd
         import matplotlib.pyplot as plt
         import seaborn as sns
         import matplotlib.pyplot as plt
         import pandas as pd
         # importing the modules
         from bokeh.plotting import figure, output_file, show
         from bokeh.palettes import magma
         import pandas as pd
```

```
In [2]:  data = pd.read_csv (r'C:\Users\nitin\Desktop\MRFL.csv')
         print (data)
```

	Date	Open	High	Low	Close \
0	1/2/2023	88600.00000	88745.35156	87879.70313	88051.20313
1	1/3/2023	88397.89844	89121.00000	87962.70313	88876.14844
2	1/4/2023	88892.00000	89073.54688	87398.00000	88012.25000
3	1/5/2023	88500.00000	91900.00000	88441.95313	91275.79688
4	1/6/2023	91580.00000	93400.00000	90557.79688	93141.50000
5	1/9/2023	93600.00000	93998.75000	92902.00000	93447.60156
6	1/10/2023	93250.00000	94480.00000	93093.89844	94163.95313
7	1/11/2023	94261.00000	94399.85156	90750.00000	91142.00000
8	1/12/2023	91597.75000	91663.50000	88573.00000	89518.25000
9	1/13/2023	89700.00000	90650.00000	89156.29688	89712.95313
10	1/16/2023	90100.00000	90679.60156	89000.00000	89228.35156
11	1/17/2023	89682.85156	89965.89844	89010.04688	89526.75000
12	1/18/2023	89360.00000	90399.95313	89279.95313	90196.85156
13	1/19/2023	89999.00000	90899.00000	89819.00000	90484.00000
14	1/20/2023	90484.00000	91198.89844	89255.64844	89557.00000
15	1/23/2023	89995.00000	90539.75000	89679.95313	90280.35156
16	1/24/2023	90500.00000	90949.39844	90045.04688	90414.50000
17	1/25/2023	90300.00000	90436.00000	88429.60156	89571.75000

Figure 13.1 Creating a data frame.

13.2.1 Data Visualization Using Scatter Plot

It is applied to find out the relationship between two variables (Refer Figure 13.2). It is used to find out correlation and autocorrelation lag in regression and time series panel data analysis. We have taken the opening price and closing price of the MRF stock to find out the relationship between the two variables. The scatter plot shows a close

```
In [3]:   # Scatter plot with OPEN against CLOSE
          plt.scatter(data['Open'], data['Close'])

          # Adding Title to the Plot
          plt.title("Scatter Plot")

          # Setting the X and Y Labels
          plt.xlabel('Open')
          plt.ylabel('Close')

          plt.show()
```

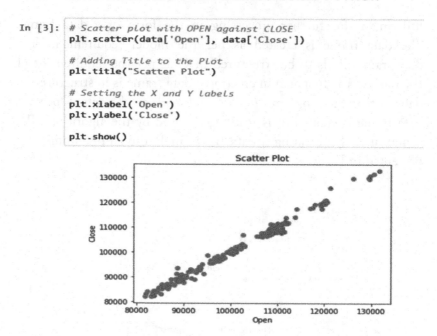

Figure 13.2 Creating a scatter plot for understanding the degree of association between the closing price and the opening price.

degree of association between them, and hence we can conclude that the opening price and closing price of the MRF stock remain the same.

It is applied to find out the relationship between two variables. It is used to find out correlation and autocorrelation lag in regression and time series panel data analysis. We have taken the opening price of MRF stock and the closing price of stock to find out the relationship between the two variables the scatter plot shows a close degree of association hence we can conclude that the opening price and closing price of MRF stock remain the same.

13.3 Data Visualization Using Bar Chat

A bar chart represents data with rectangular bars on the axis with lengths and heights that are proportional to the variable's value (Refer Figure 13.4a). It is created using the bar method. We have applied a histogram to analyze the daily movement of the opening price of the stock (Refer Figure 13.3).

```
In [4]: # Scatter plot
        plt.plot(data['Open'])
        plt.plot(data['Close'])

        # Adding Title to the Plot
        plt.title("Scatter Plot")

        # Setting the X and Y Labels
        plt.xlabel('High')
        plt.ylabel('Low')

        plt.show()
```

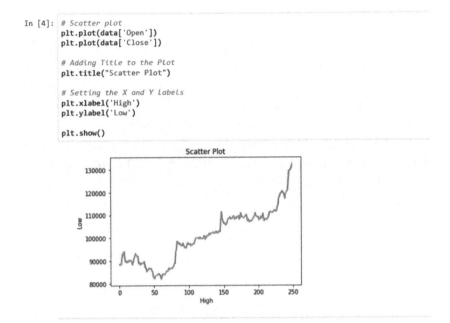

Figure 13.3 Creating Scatter Plot for understanding degree of association between closing price and opening price.

```
In [6]: # histogram
        plt.hist(data['Open'])

        plt.title("Histogram")

        # Adding the Legends
        plt.show()
```

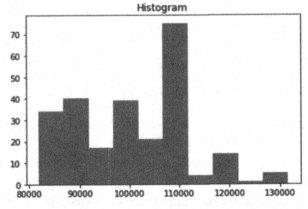

Figure 13.4a Creating a histogram for understanding the movement of the opening price.

13.4 Data Visualization Using Line Chart

It is applied to find out the relationship between two variables (Refer Figure 13.4b). It is used to find out correlation and autocorrelation lag in regression and time series panel data analysis. We have taken the opening price and closing price of the MRF stock to find out the relationship between the two variables. The scatter plot shows a close degree of association between them and hence we can conclude that the opening price and closing price of the MRF stock remain the same.

```
In [7]:  # draw lineplot
         sns.lineplot(x="Open", y="Close", data=data)

         # setting the title using Matplotlib
         plt.title('Title using Matplotlib Function')

         plt.show()
```

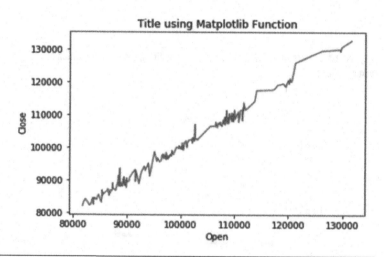

Figure 13.4b Creating a line plot for understanding the degree of association between the closing price and the opening price.

13.5 Data Visualization Using Bokeh

It generates interactive charts by generating HTML java script which uses web browsers (Refer Figure 13.5). It has a very high level of interactivity. It is applied to find out the relationship between two variables. It is used to find out correlation and autocorrelation lag in regression and time series panel data analysis. We have taken the opening price and closing price of the MRF stock to find out the

```
In [8]:   # instantiating the figure object
          graph = figure(title = "MRFL")

          color = magma(256)

          # plotting the graph
          graph.scatter(data['Open'], data['Close'], color=color)

          # displaying the model
          show(graph)
```

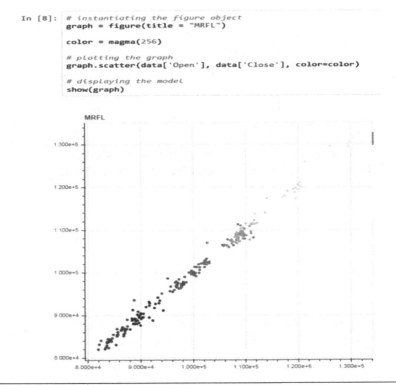

Figure 13.5 Creating scatter graph for understanding the degree of association between the closing price and the opening price.

relationship between the two variables. The scatter plot shows a close degree of association between them and hence we can conclude that the opening price and closing price of the MRF stock remain the same.

References

Hunter, J. D. (2007). Matplotlib: A 2D graphics environment. Computing in Science & Engineering, 9(3), 90–95.

Jones, E., Oliphant, T., & Peterson, P. (2019). SciPy: Open source scientific tools for Python.

McKinney, W. (2017). Python for Data Analysis: Data Wrangling with Pandas, NumPy, and IPython. O'Reilly Media, Inc.

Smith, J. (2018). Python Data Visualization Cookbook. Packt Publishing Ltd.

VanderPlas, J. T. (2016). Python Data Science Handbook: Essential Tools for Working with Data. O'Reilly Media, Inc.

Virtanen, P., Gommers, R., Oliphant, T. E., Haberland, M., Reddy, T., Cournapeau, D., . . . & van der Walt, S. J. (2020). SciPy 1.0: Fundamental algorithms for scientific computing in Python. Nature Methods, 17(3), 261–272.

Wang, J., & Liu, S. (2020). Python for Finance Cookbook. Packt Publishing Ltd.

Waskom, M., Botvinnik, O., O'Kane, D., Hobson, P., Ostblom, J., Lukauskas, S., . . . & Halchenko, Y. (2020). Mwaskom/Seaborn: v0.11.1 (December 2020). Zenodo.

Wickham, H., & Grolemund, G. (2017). R for Data Science: Import, Tidy, Transform, Visualize, and Model Data. O'Reilly Media, Inc.

14

APPLYING NATURAL LANGUAGE PROCESSING FOR STOCK INVESTORS SENTIMENT ANALYSIS

14.1 Introduction

The natural language processing (NLP) technique is used to understand the sentiments related to buying a particular stock. NLP is a technique through which we can understand sentiments by analyzing the text or comments posted on social media like Twitter, Facebook, etc. NLP classifies the text into positive and negative sentiments, which can measure the sentiments related to stock investors. NLP is a tool that can classify text and further utilize it to understand the sentiments. Hence to understand the importance of NLP, we conducted the study titled Applying Natural Language Processing for Stock Investors Sentiment Analysis (Al-Rfou & Perozzi, 2019; Bengfort et al., 2018; Bird et al., 2009).

Bird et al. (2009) applied NLP in a simple and lucid manner and his work is a big and rich resource for researchers. Jurafsky and Martin (2019) developed a conceptual literature on NLP. Manning and Schütze (1999) developed a statistical NLP method, which is the landmark work in the area of NLP. Raschka and Mirjalili (2019) give details on implementation of NLP with machine learning models. Vaswani et al. (2017) and Perkins (2016) applied the NLTK library for text mining. Loper and Bird (2002) and Chollet (2018) applied a deep planning model with AI. AI-Rfou & Perozzi (2019) analyzed the multilingual problems in NLP. Vaswani et al. (2017) developed sequence to sequence learning (Chollet, 2018).

14.2 Research Methodology

14.2.1 Data Source

Data was collected from the WhatsApp chat investors group.

14.2.2 Period of Study

The study period was from 1 January 2023 to 1 February 2023.

14.2.3 Software Used for Data Analysis

Python Programming, Anaconda

14.2.4 Model Applied

For this study, we applied the natural language processing technique.

14.2.5 Limitations of the Study

The study is restricted to natural language processing.

14.2.6 Future Scope of the Study

In the future, the study can be conducted on different stocks at the same time.

14.3 Fetching the Data into a Python Environment

Raw data filtering is a procedure applied in feature engineering (Refer Figure 14.1). Feature engineering is a process of converting raw data into features that can be utilized in an ML model. The data was fetched in the Python Anaconda environment using the Jupyter Notebook as the format of the data file was not readable in Python. A data frame is created by fetching the comma-separated values (CSV) file, making it readable in Python and utilizing it for further processing in the form of a data frame. The data frame is created as per the model require-ment. The data frame needs to be structured in as per the requirement of

```
In [1]: import pandas as pd
        import numpy as np
        import seaborn as sns
        import matplotlib.pyplot as plt
        import re
        import nltk
        nltk.download('stopwords')
        from nltk.corpus import stopwords

        [nltk_data] Downloading package stopwords to
        [nltk_data]     C:\Users\nitin\AppData\Roaming\nltk_data...
        [nltk_data]   Package stopwords is already up-to-date!

In [2]: import pandas as pd
        data = pd.read_csv (r'C:\Users\nitin\Desktop\stockremark.csv'
        print (data)
```

	OriginalTweet		Sentiment
0	no	Extremely	Negative
1	not possible	Extremely	Negative
2	never	Extremely	Negative
3	no way	Extremely	Negative
4	no @	Extremely	Negative
5	never	Extremely	Negative
6	no #	Extremely	Negative
7	no #	Extremely	Negative
8	not interested		Negative
9	no money		Negative
10	no interest		Negative
11	zero interested		Negative
12	no chance		Negative
13	no chance		Negative
14	yes		Positive
15	yes we can		Positive
16	yes we will		Positive
17	yes we can		Positive
18	will do		Positive
19	will sure		Positive
20	sure		Positive
21	yes we can		Positive
22	yes		Positive
23	yes		Positive
24	yes @		Positive
25	sure #		Positive
26	sure #		Positive
27	yes sure #		Positive
28	will do it #		Positive
29	yes sure #		Positive
30	yes will do		Positive
31	sure yes		Positive
32	will yes sure		Positive
33	done	Extremely	Positive
34	yes done	Extremely	Positive
35	# done	Extremely	Positive
36	done @	Extremely	Positive
37	@ done	Extremely	Positive
38	done @	Extremely	Positive
39	can not say		Neutral
40	can not say		Neutral
41	@can not say		Neutral
42	can not say		Neutral
43	can not say		Neutral
44	will think		Neutral
45	can not say @ will think		Neutral
46	will think		Neutral
47	not say		Neutral
48	not say		Neutral

Figure 14.1   Python libraries for performing NLP and fetching data sets into a Python environment.

the model. The first step in creating a data frame is to structure the data so that the program can read and work on the data. Once the data frame is created, it is ready to be used by the algorithm. The syntax used for creating a data frame in Python Programming is presented in Figure 14.1.

14.4 Sentiments Count for Understanding Investors' Perceptions

The sentiments are nothing but the perception of investors regarding the investment (Refer Figure 14.2). The investor's sentiments may be positive or negative depending upon their choices. We need to consider and understand investors' perceptions as they play an important role in deciding the future sales and marketing plan for selling a financial product. To understand the gravity of investors' sentiments, we need to count the number of positive and negative sentiments and analyze it (Refer Figure 14.2).

```
In [3]:  df = pd.DataFrame(data)
         df.head()

         plt.figure(figsize=(10,5))
         sns.countplot(x='Sentiment', data=df, order=['Extremely Negative', 'Negative', 'Neutral', 'Positive', 'Extremely Positive'], )
```

Out[3]: <matplotlib.axes._subplots.AxesSubplot at 0xb7b9080>

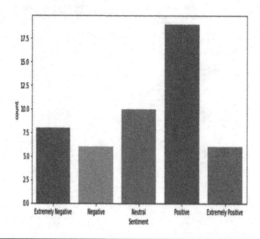

Figure 14.2 Count of sentiments.

14.5 Performing Data Cleaning in Python

The first step in natural language processing is the removal of unnecessary words (Refer Figure 14.3). The process of cleaning the text data is known as normalization. Normalization is the first step in natural language processing. Natural language processing cleans the text data by punctuation removal, stop word removal, stemming, and lemmatization. The process of normalization of data in the form of text starts with case normalization. Under case normalization, the uppercase word is converted into lowercase to standardize the overall text available for creating an NLP model. After case normalization, we step into punctuation removal. At this stage, we remove special characters and punctuation marks from the text data, making it easy for further analysis. After the punctuation removal, we apply the stop word removal technique to remove similar words. Then, we apply stemming, a process through which the fixes and prefixes of particular words are removed to normalize the given text data.

```
In [5]: reg = re.compile("(@[A-Za-z0-9]+)|(#[A-Za-z0-9]+)|([^0-9A-Za-z t])|(w+://S+)")
        tweet = []
        for i in df["OriginalTweet"]:
          tweet.append(reg.sub(" ", i))
        df = pd.concat([df, pd.DataFrame(tweet, columns=["CleanedTweet"])], axis=1, sort=False)
```

```
In [6]: df.head()
```

Out[6]:

	OriginalTweet	Sentiment	CleanedTweet
0	no	Extremely Negative	no
1	not possible	Extremely Negative	not possible
2	never	Extremely Negative	never
3	no way	Extremely Negative	no way
4	no @	Extremely Negative	no

```
In [7]: from sklearn.feature_extraction.text import TfidfVectorizer
        stop_words = set(stopwords.words('english'))    # make a set of stopwords
        vectoriser = TfidfVectorizer(stop_words=None)
```

Figure 14.3   Performing data cleaning in Python.

14.6 Performing Vectorization in Python

In natural language processing, we convert the textual data into numerical values that can be easily understood by the machine learning algorithm (Refer Figure 14.4). The neural language process allows computers to understand the process of human language by understanding meaning and context related to that particular sentiment, through which we can get a useful inside in creating and understanding NLP. NLP combines computational linguistics with statistical machine learning algorithms through which the model is created for analysis purposes. Vectorization is a classic approach to converting linguistic text into real numbers in the form of vectors which are applied for supporting and creating a machine learning model. The vectorization is the first step in the extraction of features. The basic idea behind vectorization is to get distinct features out of the linguistic text available for analysis. The last step is to create a training and test model for analysis.

14.7 Vector Transformation to Create Trial and Training Data Sets

After cleaning the data for sentiment analysis, we need to create trial and training data sets (Refer Figure 14.5). To test the accuracy of the

```
In [8]: X_train = vectoriser.fit_transform(df["CleanedTweet"])
        # Encoding the classes in numerical values
        from sklearn.preprocessing import LabelEncoder
        encoder = LabelEncoder()
        y_train = encoder.fit_transform(df['Sentiment'])
        from sklearn.naive_bayes import MultinomialNB
        classifier = MultinomialNB()
        classifier.fit(X_train, y_train)
```

Out[8]: MultinomialNB(alpha=1.0, class_prior=None, fit_prior=True)

Figure 14.4 Performing Vectorization in Python.

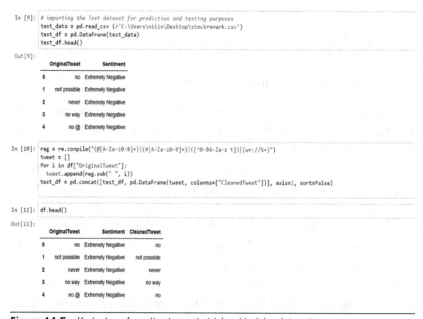

Figure 14.5 Vector transformation to create trial and training data sets.

model applied, we need to test the results by comparing it with original data (Refer Figure 14.4). For comparing the data sets, we need to divide the data set into test and train. A label encoder is used to transform the data of sentiments into numerical values for creating a multinomial Naive Bayes model. The process of transforming the sentiments from text to numerical values is known as vector transformation.

14.8 Result Analysis Model Testing AUC

The AUC stands for the area under the ROC curve (Refer Figure 14.6). The AUC provides the accuracy of all possible classifications with its thresholds. The AUC is the ratio between the true positive rate and the false positive rate. An AUC value of 1 is considered to be the best with 100 percent model accuracy. The value of AUC ranges from 0 to 1. In the present study, we have an AUC value of 0.20, which is considered to be poor.

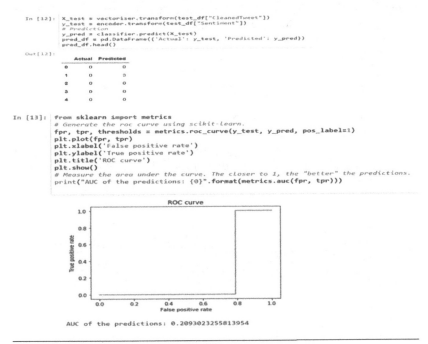

Figure 14.6   Result analysis model testing AUC.

14.9 Conclusion

The implementation of natural language processing requires a huge amount of critical analysis. After the implementation of natural language processing, we implemented the Naive Bayes model, which has a low accuracy of 20 percent.

References

Al-Rfou, R., & Perozzi, B. (2019). Polyglot: Distributed Word Representations for Multilingual NLP. Proceedings of the 57th Annual Meeting of the Association for Computational Linguistics.

Bengfort, B., Bilbro, R., & Ojeda, T. (2018). Applied Text Analysis with Python: Enabling Language-Aware Data Products with Machine Learning. O'Reilly Media.

Bird, S., Klein, E., & Loper, E. (2009). Natural Language Processing with Python. O'Reilly Media.

Chollet, F. (2018). Deep Learning with Python. Manning Publications.

Jurafsky, D., & Martin, J. H. (2019). Speech and Language Processing (3rd ed.). Pearson.

Loper, E., & Bird, S. (2002). NLTK: The Natural Language Toolkit. CoRR. cs.CL/0205028. 10.3115/1118108.1118117.

Manning, C. D., & Schütze, H. (1999). Foundations of Statistical Natural Language Processing. The MIT Press.

Perkins, J. (2016). Python Text Processing with NLTK 2.0 Cookbook. Packt Publishing.

Raschka, S., & Mirjalili, V. (2019). Python Machine Learning: Machine Learning and Deep Learning with Python, scikit-learn, and TensorFlow (3rd ed.). Packt Publishing.

Ashish Vaswani, Noam Shazeer, Niki Parmar, Jakob Uszkoreit, Llion Jones, Aidan N. Gomez, Łukasz Kaiser, and Illia Polosukhin. 2017. Attention is all you need. In Proceedings of the 31st International Conference on Neural Information Processing Systems (NIPS'17). Curran Associates Inc., Red Hook, NY, USA, 6000–6010.

15

STOCK PREDICTION
APPLYING LSTM

15.1 Introduction

Stock prediction and modeling involve a huge amount of analysis. The analysis includes different models applied for analysis. It includes the logistic regression model, regression analysis, and support vector machines, but all these models involve analysis of data that is not panel data. For panel data analysis, we use models like autoregressive moving average (ARIMA) and LSTM. In recent times, the LSTM model, which is popularly known as long short-term memory, has had a huge impact on the prediction of stocks as applied to time series data. LSTM is a deep learning model that is based on the principle of neural networking. The LSTM model consists of three layers: input layer, hidden layer, and output layer. The capability of memorizing the sequence of data makes LSTM a specialized type of recurrent neural network, and hence the study titled Stock Prediction by Applying LSTM was conducted in three steps: fetching the data, cleaning the data number, data transformation and normalization, model analysis.

The LSTM model is known for predicting and developing a predictive model that involves long short-term memory and the removal of irrelevant information. The irrelevant information is deleted from the model in the form of a forget gate. For example, we explain this with the sentence Dr Nitin Stays at Aurangabad (Refer Table 15.1). Dr Nitin has an area of specialization in Financial Analytics. He is a good teacher in the area of _____. From the analysis of the above sentence, we need to fill up the blank space with relevant information. Here the sentence includes information about Dr Nitin and the area of specialization in which he works can be used to fill up the blank space. To fill up the blank space, we need to analyze the relevant information and after analyzing the information from the sentence, the relevant and

Table 15.1 The Architecture and Process of the Long Short-Term Model

LSTM PROCESS	SENTENCE
Long-term memory	Dr Nitin stays at Aurangabad. Dr Nitin has an area of specialization in Financial Analytics. He is a good teacher in the area of_____
Short-term memory	Dr. Nitin has an area of specialization in Financial Analytics. He is a good teacher in the area of_____
Input information	Dr Nitin stays at Aurangabad. Dr. Nitin has an area of specialization in Financial Analytics. He is a good teacher in the area of_____
Irrelevant information or Forget information	Dr. Nitin stays at Aurangabad
Relevant information	Area of specialization in Financial Analytics.
Output	He is a good teacher in the area of **Financial Analytics**.

irrelevant information produced by analyzing it is required to be used for filling up the blank space. Irrelevant information is to be removed from the analysis. The irrelevant information is about the residence of Dr Nitin which needs to be removed (Refer Table 15.1). The LSTM model that we apply here includes long-term memory in the form of the overall sentence. After removing the irrelevant information, the remaining information is the short-term memory. Here after further analysis, we need to forget about the information regarding the residence of Dr Nitin and we need to analyze the information about the area of specialization in which Dr Nitin works. After applying the past knowledge and long short-term memory we delete the irrelevant information and the blank space is filled up with the correct answer. The correct answer for filling up the blank space is Financial Analytics. Table 15.1 explains the long short-term model.

15.1.1 Review of Literature

The LSTM model was developed by Hochreiter and Schmidhuber (1997). The model is based on recurrent neural network to overcome vanishing gradient problems (Gers et al., 1999; Graves et al., 2013, 2014). These models are based on sequential data which helps in developing speech recognition (Graves et al., 2013). LSTM models have network units where information moves in a regulated manner (Greff et al., 2016; Sundermeyer et al., 2012). Due to gradient problems, it is difficult to train LSTM models (Pascanu et al., 2013). The bidirectional LSTM architecture was proposed by Schuster and Paliwal (1997) and neural turing machines were developed by Graves et al. (2014; Chen et al., 2016; Lipton 2015).

15.2 Research Methodology

15.2.1 Data Source

Data was collected from Yahoo Finance.

15.2.2 Period of Study

The study period was from 1 January 2023 to 1 February 2023.

15.2.3 Software Used for Data Analysis

Python Programming, Anaconda

15.2.4 Model Applied

For this study, we applied the long short-term memory (LSTM) model.

15.2.5 Limitations of the Study

The study is restricted to the long short-term memory (LSTM).

15.2.6 Future Scope of the Study

In the future, the study can be conducted on different stocks at the same time.

15.3 Fetching the Data into a Python Environment

Raw data filtering is a procedure applied in feature engineering (Refer Figure 15.1). Feature engineering is a process of converting raw data into features that can be utilized in an ML model. The data was fetched in the Python Anaconda environment using the Jupyter Notebook as the format of the data file was not readable in Python. A data

```
In [1]:  # Make sure that you have all these libaries available to run the code successfully

         import matplotlib.pyplot as plt
         import pandas as pd
         import datetime as dt
         import urllib.request, json
         import os
         import numpy as np
         import tensorflow as tf # This code has been tested with TensorFlow 1.6
         from sklearn.preprocessing import MinMaxScaler
         from keras.models import Sequential
         from keras.layers import Dense, Dropout, LSTM, Bidirectional
```

```
In [2]:  gstock_data = pd.read_csv (r'C:\Users\Ishan\OneDrive\Desktop\MRF.csv')
         gstock_data .head()
```

Out[2]:

	Date	Open	Close
0	23-03-2023	84949.95313	84168.39844
1	24-03-2023	84046.39844	83871.79688
2	27-03-2023	84299.60156	83602.95313
3	28-03-2023	84150.00000	82342.45313
4	29-03-2023	82005.00000	81982.60156

Figure 15.1   Python libraries for performing LSTM and fetching data sets into a Python environment.

frame is created by fetching the comma-separated values (CSV) file, making it readable in Python and utilizing it for further processing in the form of a data frame. The data frame is created as per the model requirement. The data frame needs to be structured as per the requirement of the model. The first step in creating a data frame is to structure the data so that the program can read and work on the data. Once the data frame is created, it is ready to be used by the algorithm. The syntax used for creating a data frame in Python Programming is presented in Figure 15.1.

15.4 Performing Data Cleaning in Python

The data frame is prepared to maintain the notion of the model which has different variables (Refer Figure 15.2). The Feature engineering is the process of preparing converting the data into nominal scale or ordinal scale etc. it prepare data that can be read and utilized by algorithm. data which will be ready for the program to utilize. The syntax used for creating a data frame in Python Programming is presented in Figure 15.2.

```
In [3]:  gstock_data = gstock_data [['Date','Open','Close']]
         gstock_data .set_index('Date',drop=True,inplace=True)
         gstock_data .head()
```

```
Out[3]:
                         Open        Close

              Date

         23-03-2023   84949.95313   84168.39844

         24-03-2023   84046.39844   83871.79688

         27-03-2023   84299.60156   83602.95313

         28-03-2023   84150.00000   82342.45313

         29-03-2023   82005.00000   81982.60156
```

Figure 15.2 Performing data cleaning in Python.

15.5 Vector Transformation to Create Trial and Training Data Sets

After cleaning the data for the LSTM ML model, we need to create trial and testing data sets (Refer Figure 15.3). To test the accuracy of the model applied, we need to test the results by comparing it with original data. For comparing the data sets, we need to divide the data set into test and train. For model evaluation, we divide the data into 80 percent of trial data and 20 percent of test data by vector transformation.

```
In [4]:  from sklearn.preprocessing import MinMaxScaler
         Ms = MinMaxScaler()
         gstock_data [gstock_data .columns] = Ms.fit_transform(gstock_data )
         training_size = round(len(gstock_data ) * 0.80)
         train_data = gstock_data [:training_size]
         test_data = gstock_data [training_size:]
```

```
In [5]:  train_seq = train_data [:training_size]
         create_sequence = test_data [training_size:]
```

Figure 15.3 Vector transformation to create trial and training data sets.

15.6 Result Analysis for the LSTM Model

The results of the LSTM model show the different layers in the output and the generated parameters for the different layers (Refer Figure 15.4). We applied LSTM layers of 50 units. The representation of parameters in LSTM units are functions involved in calculations. The parameters are generated using the below formula:

$$\text{Number of parameters} = 4 * (N + M + 1) * m$$

where
N is the number of dimensions in the input variable.
M is the units in the LSTM layer.
1 is the bias parameter.

Substituting the values in the above equation, we get

$$\text{Number of LSTM parameters} = 4 * (50 + 50 + 1) * 50$$
$$\text{Number of LSTM parameters} = 20,200$$

We applied the LSTM model and generated different layers for predicting the opening price and the closing price. The LSTM model is generated with 50 units or neurons. These units from the LSTM

```
In [6]: model = Sequential()
        model.add(LSTM(units=50, return_sequences=True, input_shape = (train_seq.shape[0], train_seq.shape[1])))

        model.add(Dropout(0.1))
        model.add(LSTM(units=50))

        model.add(Dense(2))

        model.compile(loss='mean_squared_error', optimizer='adam', metrics=['mean_absolute_error'])

        model.summary()
```

Model: "sequential"

Layer (type)	Output Shape	Param #
lstm (LSTM)	(None, 197, 50)	10,600
dropout (Dropout)	(None, 197, 50)	0
lstm_1 (LSTM)	(None, 50)	20,200
dense (Dense)	(None, 2)	102

Total params: 30,902 (120.71 KB)
Trainable params: 30,902 (120.71 KB)
Non-trainable params: 0 (0.00 B)

Figure 15.4   Result analysis for the LSTM model.

model will be used as input to the next LSTM layer. The next dropout layer is the regulator of the model, which keeps the irrelevant information or the so-called biases away from the LSTM model. The other LSTM-1 layer with 50 neurons or units is followed by the final dense LSTM layer with 2 neurons or units.

15.7 Conclusion

The long short-term memory model is created for predicting the opening price and the closing price. The total parameters generated for the LSTM layer is 20,200. After applying the past knowledge and the LSTM, we deleted the irrelevant information and created an LSTM model for predicting the stock price.

References

Chen, X., Duan, Y., Houthooft, R., Schulman, J., Sutskever, I., & Abbeel, P. (2016). Infogan: Interpretable representation learning by information maximizing generative adversarial nets. In Advances in Neural Information Processing Systems (pp. 2172–2180). https://api.semanticscholar.org/CorpusID:5002792

Gers, F. A., Schmidhuber, J., & Cummins, F. (1999). Learning to forget: Continual prediction with LSTM. Neural Computation, 12(10), 2451–2471.

Graves, A., Mohamed, A., & Hinton, G.E. (2013). Speech recognition with deep recurrent neural networks. 2013 IEEE International Conference on Acoustics, Speech and Signal Processing, 6645–6649.

Graves, A., Wayne, G., & Danihelka, I. (2014). Neural turing machines. arXiv preprint arXiv:1410.5401.

Greff, K., Srivastava, R. K., Koutník, J., Steunebrink, B. R., & Schmidhuber, J. (2016). LSTM: A search space odyssey. IEEE Transactions on Neural Networks and Learning Systems, 28(10), 2222–2232.

Hochreiter, S., & Schmidhuber, J. (1997). Long short-term memory. Neural Computation, 9(8), 1735–1780.

Lipton, Z. C., Berkowitz, J., & Elkan, C. (2015). A critical review of recurrent neural networks for sequence learning. arXiv preprint arXiv:1506.00019.

Pascanu, R., Mikolov, T., & Bengio, Y. (2013). On the difficulty of training recurrent neural networks. In International Conference on Machine Learning (pp. 1310–1318).

Schuster, M., & Paliwal, K. K. (1997). Bidirectional recurrent neural networks. IEEE Transactions on Signal Processing, 45(11), 2673–2681.

Sundermeyer, M., Schlüter, R., & Ney, H. (2012). LSTM neural networks for language modeling. In Thirteenth Annual Conference of the International Speech Communication Association.

Printed in the United States
by Baker & Taylor Publisher Services